Texts and Monographs in Computer Science

The AKM Series in
Theoretical Computer Science

A Subseries of Texts and Monographs in Computer Science

A Basis for Theoretical Computer Science
by M. A. Arbib, A. J. Kfoury, and R. N. Moll

A Programming Approach to Computability
by A. J. Kfoury, R. N. Moll, and M. A. Arbib

An Introduction to Formal Language Theory
by R. N. Moll, M. A. Arbib, and A. J. Kfoury

(iii) The set $\{\Lambda\}$ whose only element is the empty string is an FSL, since it is accepted by the machine.

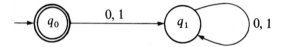

Here $F = \{q_0\}$ and we see that $\delta^*(q_0, w) = q_0$ if $w = \Lambda$, while $\delta^*(q_0, w) = q_1$ if $w \neq \Lambda$. Then $T(M) = \{w \mid \delta^*(q_0, w) = q_0\} = \{\Lambda\}$.

(iv) The string 1011 (i.e., the one-element subset $\{1011\}$ of X^*) is accepted by the machine

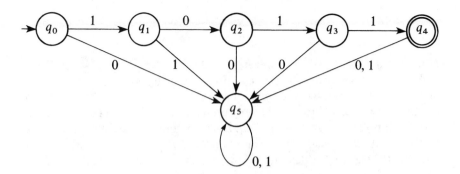

The state q_5 (like the state q_1 of the previous example) is a *trap* — it is a nonaccepting state with the property that once M enters it, M never leaves it: $\delta(q_5, x) = q_5$ for each x in X. We may simplify state diagrams by omitting all trap states and the edges which lead to them. Thus the above state graph can be abbreviated to

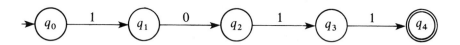

it being understood that each missing transition leads to a single trap state. With this convention, we see that any string $x_1 x_2 \cdots x_n$ $(n \geq 0)$ is accepted by an $(n + 2)$-state machine

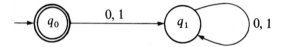

(Why does this diagram only show $n + 1$ states? Does this work for $n = 0$?)

(v) We can extend this construction to accept any finite subset of X^*. First, we show the construction for $L = \{0, 11, 101\}$:

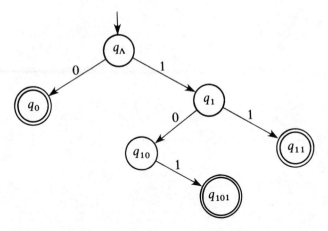

Note that q_Λ, *not* q_0, is the initial state in this diagram.

Here is the general construction. For any string $w = x_1 \cdots x_n$ $(n \geq 0)$ of X^*, we say a string w' is a prefix of w just in case $w' = x_1 \cdots x_m$ for some m with $0 \leq m \leq n$. (Thus Λ is the only prefix of Λ; each w is a prefix of itself; and Λ is a prefix of any w.) Let, then, L be any finite subset of X^* and let K be the set of all prefixes of L. (For example, if $L = \{0, 11, 101\}$, then $K = \{\Lambda, 0, 1, 10, 11, 101\}$.) Then an FSA M_L which satisfies $L = T(M_L)$ may be defined as follows:

The set Q_L of states has one distinct state q_w for each string w in the set K of prefixes of L, together with a trap state q_t. The initial state is q_Λ. For each w in K and x in X

$$\delta_L(q_w, x) = \begin{cases} q_{wx}, & \text{if } wx \text{ is in } K, \\ q_t, & \text{if not,} \end{cases}$$

while $\delta_L(q_t, x) = q_t$ for each x in X.

Finally, $F_L = \{q_w | w \text{ is in } L\}$.

This reduces to the construction in (v) above for $L = \{0, 11, 101\}$, as well as (except for renaming of the states) that of (iv) for $L = \{1011\}$, that of (iii) for $L = \{\Lambda\}$, and that of (i) for $L = \emptyset$. (Check the construction for each of these cases to make sure you agree.)

6 Proposition. *Each finite subset of X^* is a finite-state language.* ☐

EXERCISES FOR SECTION 2.3

1. Give FSAs which accept each of the following languages.
   (i) $L = \{001, 100, 11\}$
   (ii) $L = \{1^k | k \text{ is divisible by 3}\}$
   (iii) $L = \{1^k 0^j | k \text{ is divisible by 3, and } j \text{ is odd}\}$.

A Basis
for Theoretical
Computer Science

Michael A. Arbib
A. J. Kfoury
Robert N. Moll

With 49 Figures

Springer-Verlag
New York   Heidelberg   Berlin

Michael A. Arbib
Department of Computer and
    Information Science
University of Massachusetts
Amherst, MA 01003
USA

A. J. Kfoury
Department of Mathematics
Boston University
Boston, MA 02215
USA

Series Editor

Robert N. Moll
Department of Computer and
    Information Science
University of Massachusetts
Amherst, MA 01003
USA

David Gries
Department of Computer Science
Cornell University
Upson Hall
Ithaca, NY 14859
USA

AMS Subject Classification (1980): 68C01 68D05 68F05
(C.R.) Computer Classification: 5.20

With 49 Figures

Library of Congress Cataloging in Publication Data
Arbib, Michael A.
  A basis for theoretical computer science.

  (Texts and monographs in computer science)
  Bibliography: p.
  Includes index.
  1. Machine theory.   2. Formal languages.   I. Kfoury, A. J.
  II. Moll, Robert N.   III. Title.   IV. Series.
QA267.A715   001.64   81-5688
                    AACR2

© 1981 by Springer-Verlag New York Inc.
Printed in the United States of America.

9 8 7 6 5 4 3

ISBN 0-387-90573-1   Springer-Verlag   New York   Heidelberg   Berlin
ISBN 3-540-90573-1   Springer-Verlag   Berlin   Heidelberg   New York

Preface

Computer science seeks to provide a scientific basis for the study of information processing, the solution of problems by algorithms, and the design and programming of computers. The last forty years have seen increasing sophistication in the science, in the microelectronics which has made machines of staggering complexity economically feasible, in the advances in programming methodology which allow immense programs to be designed with increasing speed and reduced error, and in the development of mathematical techniques to allow the rigorous specification of program, process, and machine. The present volume is one of a series, The AKM Series in Theoretical Computer Science, designed to make key mathematical developments in computer science readily accessible to undergraduate and beginning graduate students. Specifically, this volume takes readers with little or no mathematical background beyond high school algebra, and gives them a taste of a number of topics in theoretical computer science while laying the mathematical foundation for the later, more detailed, study of such topics as formal language theory, computability theory, programming language semantics, and the study of program verification and correctness.

Chapter 1 introduces the basic concepts of set theory, with special emphasis on functions and relations, using a simple algorithm to provide motivation. Chapter 2 presents the notion of inductive proof and gives the reader a good grasp on one of the most important notions of computer science: the recursive definition of functions and data structures. Chapter 2 also introduces the reader to formal language theory, and to list processing. Chapter 3 examines trees, structures which recur again and again in computer science, and shows how techniques for counting trees enable us to solve a whole range of interesting problems. We also give an example of "analysis of algorithms"—showing how counting the number of execution steps of an algorithm allows us to compare the efficiency of different approaches to a given problem.

Chapter 4 looks at the role of the two element set in the analysis of switching circuits and in the proving of theorems. Chapter 5 gives us a more detailed study of the relations of Chapter 1, with special emphasis on equivalence relations and partial order. We introduce lattices and Boolean algebras as classes of partially ordered sets of special interest to computer scientists. Finally we introduce Cantor's diagonal argument which not only shows that infinities come in different sizes, but also plays a vital role in computability theory. The last Chapter is an introduction to graph theory, motivated by a study of Euler's 1735 study of "The Seven Bridges of Königsberg." We use matrices over semirings to study the connectivity of graphs and the reachability problem for automata, and close by proving Kleene's theorem on the equivalence of finite-state languages and regular languages. More detailed outlines may be found at the start of each section.

The book grew out of our teaching to classes over several years at the University of Massachusetts at Amherst. Our colleague, Edwina Rissland, taught from an earlier draft of the book and provided us with many constructive suggestions embodied in the published text. We thank our students for all that they taught us about what we had to teach them, and we thank Gwyn Mitchell, Martha Young and Stan Kulikowski for their efforts in typing the manuscript.

July 1980   M.A.A., A.J.K., R.N.M.

Contents

CHAPTER 5

Binary Relations, Lattices, and Infinity

CHAPTER 6

Graphs, Matrices, and Machines

CHAPTER 1

Sets, Maps, and Relations

There are two main languages in which theoretical computer science is expressed — the language of sets and the language of numbers. The principal aim of this chapter is to set forth the basic concepts of set theory. Section 1.1 introduces sets and subsets, and shows how to build up new sets from old ones by such operations as the Cartesian product, union, intersection, and set differences. Section 1.2 briefly summarizes fundamental facts about exponents, logarithms, and finite series. Section 1.3 introduces the basic notions of maps or functions, partial functions, and relations. The way in which computer programs transform input data provides examples of partial functions. Finding relations between items stored in a data base is an important task for computer science.

1.1 Sets

To motivate our introduction of set theory, we shall study a flow diagram (Figure 1) for dividing one number into another. The flow diagram accepts as input two integers, x and y (with $x \geq 0$ and $y > 0$), and finally halts after printing out two numbers r and q (r stands for "remainder"; q for "quotient.") The notation $:=$ can be read as "becomes." Thus $r := r - y$ does *not* mean that r equals $r - y$ (this would be true only if $y = 0$, which cannot be true if $y > 0$). Instead, $r := r - y$ means that we change the value of r so that it *becomes* equal to the old value of r less the old value of y. So if we start with r equal to 6, and y equal to 4, the instruction $r := r - y$ changes r to 2 (but leaves the value of y as 4).

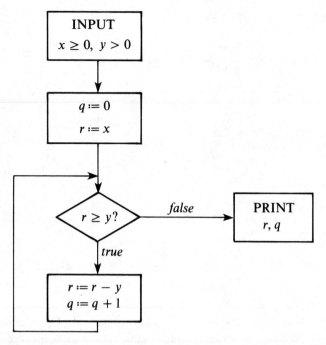

Figure 1   A simple flow diagram for computing x **mod** y and x **div** y.

After reading in its input, the program initializes r to equal x, and q to equal 0. Each loop execution decreases r by y and increases q by 1. So, after n times round the loop $q = n$ and $r = x - n * y$ (using $*$ to denote multiplication). The loop is traversed until r is less than y, so that on exit we have

$$x = q * y + r \quad \text{and} \quad 0 \le r < y$$

(can you see why $0 \le r$?), which we may rewrite as

$q = x$ **div** y,   the integer part of x divided by y; and

$r = x$ **mod** y, the remainder after x is divided by y.

SETS AND SUBSETS

We now explore some of the sets, functions, and relations that play a role in the program given in Figure 1. A set is simply a collection of objects. Sets are usually described by some specification that indicates whether or not a particular object belongs to (is a member of) the collection. The first set we meet in our example is the set of integers, which we shall write as **Z**. The program's two inputs, x and y, must be members of this set. However, by restricting x and y, we have constrained the program to deal exclusively with the set **N** of all non-negative integers — these are called the "natural numbers." We can indicate the set **N** in two ways (at least!):

$$\mathbf{N} = \{0, 1, 2, 3, \ldots\}$$

$$\mathbf{N} = \{x \,|\, x \in \mathbf{Z} \quad \text{and} \quad x \geq 0\}.$$

In the first case, we display an explicit list of members of the set **N**. In the second case, we describe conditions which must be met in order for an element to belong to the set — "**N** is the set of all x such that x is an integer and $x \geq 0$." Here we use the general notation $x \in A$ for "x is a member of the set A," which we may also read as "x is in A" or "x belongs to A." The notation $x \notin A$ indicates that x does *not* belong to A. Thus $3 \in \mathbf{N}$ but $-2 \notin \mathbf{N}$. More generally, if $P(x)$ indicates some test which may be applied to suitable x's to yield an answer of *true* or *false*, then $\{x \,|\, P(x)\}$ is the set of all x for which $P(x)$ is true. For example

$$\{x \,|\, x \in \mathbf{Z} \text{ and } x \bmod 2 = 0\} = \text{the set of even integers;}$$

$$\{x \,|\, x \in \mathbf{Z} \text{ and } x \bmod 2 = 1\} = \text{the set of odd integers;}$$

$$\{x \,|\, x \in \mathbf{Z} \text{ and } 2 \leq x < 7\} = \{2, 3, 4, 5, 6\}.$$

Another set implicit in Figure 1 is the set

$$\mathscr{B} = \{T, F\}$$

of truth values (where T is short for *true*, F for *false*). This set is also denoted **Boolean**, and is often called the set of Boolean values in honor of the British mathematician and logician George Boole (1818–1864) whose book, *An Investigation of the Laws of Thought*, published in London in 1854, showed how the truth of propositions could be manipulated in an algebraic fashion. The resultant subject of *propositional logic* and its applications in computer science will occupy us in Chapter 4. The test $r \geq y$? takes on a value that is a member of the set **Boolean**.

One important set (not explicitly mentioned in Figure 1) is the *empty set*

$$\varnothing = \{\ \}$$

which contains no elements. Just as 0 is a very useful number, so $\varnothing$ is a very useful set:

$$\varnothing = \text{the set of American unicorns}$$

$$\varnothing = \{x \,|\, x \in \mathbf{N} \text{ and } x < 0\}$$

and many other bizarre examples best left to the imagination of the reader.

We say a set A is a *subset* of the set B, and write $A \subset B$ (though some authors prefer the notation $A \subseteq B$), if every element of A is also an element of B: $x \in A$ implies $x \in B$. We say A is a *proper* subset of B if $A \subset B$ but $A \neq B$.

We note the following basic facts:

1. For any set A, both $\emptyset \subset A$ and $A \subset A$.
2. For any two sets A and B, we have that both $A \subset B$ and $B \subset A$ are true iff (short for: "if and only if") $A = B$, i.e. iff every element of A is an element of B and vice versa.

Notice that $\mathbf{N} \subset \mathbf{Z}$, and that the set of values that y can take on in our sample program — $y \in \{w \mid w > 0\}$ — is a *proper* subset of x's possible values — $x \in \{z \mid z \geq 0\}$.

We stress that the membership rule for a finite set given by an explicit list like $\{2, 3, 4, 5, 6\}$ is simply "$x \in A$ iff x is on the list." Thus the set does not change if we alter the order of elements on the list or repeat elements:

$$\{2, 3, 4, 5, 6\} = \{2, 3, 6, 5, 4\} = \{6, 2, 6, 3, 5, 4, 2, 6\}$$

but the set will change if we add or remove elements:

$$\{2, 3, 4, 5\} \neq \{2, 3, 4, 5, 6\}$$

$$\{a, 2, 3, b, 4, 5, 6\} \neq \{2, 3, 4, 5, 6\}.$$

Thus a set is different from a sequence. A sequence is a list of elements which are placed in a specific order. We use the general notation

$$(x_1, x_2, x_3, \ldots)$$

for the sequence with x_1 *before* x_2 *before* $x_3 \ldots$. We say that two sequences are equal if they contain the same elements in the same order. Thus

$$(x_1, x_2, \ldots, x_m) = (y_1, y_2, \ldots, y_n)$$

if $m = n$ (the sequences have the same *length*, namely m) and $x_1 = y_1$, $x_2 = y_2, \ldots$, and $x_m = y_n$.

We use the term n-tuple for a sequence of length n. Thus, an ordered pair is a 2-tuple.

A sequence can be finite or it might never end, in which case we say it is *infinite*. For example, the sequence of all prime numbers (see Exercise 1) in increasing order is

$$(2, 3, 5, 7, 11, 13, \ldots)$$

and the sequence never ends. But how do we know it never ends? The answer was given by Euclid, the "father of geometry," who lived in ancient Greece around 300 BC. His proof is an example of *proof by contradiction* or *reductio ad absurdum* (Latin for "reduction to an absurdity"). To prove some result P, by this method, we look at what happens if we start by assuming that P is *not* true. If this assumption leads to a contradiction, it was wrong to assume P was not true — and we conclude that P must be true after all.

Euclid's Proof That There Are Infinitely Many Primes. If there are only finitely many primes, we could write them down in increasing order as a finite sequence

$$(p_1, p_2, p_3, \ldots, p_n)$$

where $p_1 = 2$, $p_2 = 3$, $\ldots$, and p_n is the largest prime. Now consider the number obtained by multiplying all these primes together and then adding one to the result:

$$M = (p_1 * p_2 * p_3 * \cdots * p_n) + 1.$$

For any prime number p_k, we have $M \bmod p_k = 1$, so none of the primes on our list divides M evenly. Thus either M itself is a prime number not on our list, or it is a product of primes not on the list. This contradicts the assumption that there are finitely many primes. Thus the list of primes is infinite.   □

(The symbol □ is used either to mark the end of a proof, or to indicate that any further details of the proof are left to the reader.)

Given two sets A and B, we are often interested in the set, denoted $A \times B$, of all ordered pairs (a, b) whose first element a is in A and whose second element b is in B:

$$A \times B = \{(a, b) | a \in A, b \in B\}.$$

This set is called the *Cartesian product* of A and B, in honor of the French mathematician and philosopher René Descartes (1596–1650) who founded analytic geometry with the observation that the plane could (as we see in Figure 2) be represented as

$$\mathbf{R} \times \mathbf{R} = \{(x, y) | x \in \mathbf{R}, y \in \mathbf{R}\},$$

where $\mathbf{R}$ is the "real number line," that is, the set of all real numbers. The idea extends naturally to three-fold and indeed n-fold Cartesian products:

$$A \times B \times C = \{(a, b, c) | a \in A, b \in B, c \in C\}$$

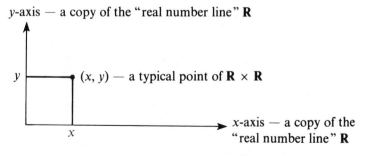

Figure 2

and

$$A_1 \times A_2 \times \cdots \times A_n = \{(a_1, a_2, \ldots, a_n) | a_1 \in A, \ldots, a_n \in A_n\}.$$

It is often useful to view a Cartesian product as a set of samples. Each sample is obtained by selecting one element from each of the sets that make up the product. Thus if $a \in A$, $b \in B$, and $c \in C$, then (a, b, c) is a "sample" from $A \times B \times C$, and b's position in this 3-tuple tells us that b came out of the second factor in the product, B.

The Cartesian product is an extremely valuable construction in mathematics and computer science. It gives us an elegant and general notation for describing a great many familiar mathematical concepts and operations.

For example, note that in Figure 1, the assignment $r := r - y$ operates on the ordered pair of integers (r, y) — the variables on the right hand side of the assignment symbol — to return the new value of r, so that we can think of this statement as "acting" on $\mathbf{N} \times \mathbf{N}$. We return to this in Section 1.3.

Let us consider another example, the set B of birthdates of people born in the 20th century. Since a birthdate is a triple of the form

(month, day, year)

we can think of B as a subset of the Cartesian product $M \times D \times Y$ where the set M of months is {January, February, ..., December}, the set D of days is $\{1, 2, \ldots, 30, 31\}$ and the set Y of years is $\{0, 1, 2, 3, \ldots, 99\}$. B is a proper subset of $M \times D \times Y$ because, for example, (February, 29, 89) does not belong to B.

Given a set A, we use $\mathscr{P}A$ or 2^A to denote the *powerset* of A, which is defined to be the set of all subsets of A.

$$\mathscr{P}A = 2^A = \{X | X \subset A\}.$$

If A is a finite set (that is, we can count how many elements there are in A), we use $|A|$ — read "the cardinality of A" — to denote the number of elements of A. For example

$$|\{2, 3, 2\}| = 2.$$

$$|\varnothing| = 0,$$

$$|\{\{a, b\}\}| = 1.$$

(Make sure you understand the point of the first and third example.) It should be easy to see that if A and B are finite sets, then $|A \times B| = |A| * |B|$.

The 0-element set $\varnothing$ has one subset:

$$\mathscr{P}\varnothing = \{\varnothing\}, \text{ so that } |\mathscr{P}\varnothing| = 2^{|\varnothing|}, \text{ since } 1 = 2^0.$$

The 1-element set $\{a\}$ has two subsets:

$$\mathscr{P}\{a\} = \{\varnothing, \{a\}\}, \text{ so that } |\mathscr{P}\{a\}| = 2^{|\{a\}|}, \text{ since } 2 = 2^1.$$

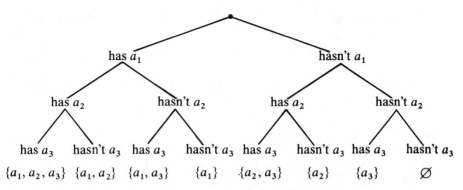

Figure 3   A "binary decision tree" for the $8 = 2^3$ subsets of $\{a_1, a_2, a_3\}$.

This suggests the general hypothesis that an n-element set has 2^n subsets. Before we give the general proof, look at Figure 3 to see how to systematically arrange the 8 subsets of the 3-element set $\{a_1, a_2, a_3\}$.

1 Fact. *If A is a finite set with m elements, then A has 2^m subsets. In symbols, we have*

$$|2^A| = 2^{|A|}$$

which explains the choice of the notation 2^A for the powerset of A.

PROOF. Suppose $A = \{a_1, a_2, \ldots, a_m\}$. Then we can divide 2^A into $2 = 2^1$ collections: subsets of A which contain a_1, and those which do not. Considering the next element a_2, we get $4 = 2^2$ collections depending on which of a_1 and a_2 are included in the subset. Continuing in this way, we finally see that there are 2^m subsets of A, each determined by the choice of whether or not each a_j will be included, as j runs from 1 to m.   □

We have already seen how to combine two sets to form their Cartesian product. We close this subsection by showing several other basic ways of combining sets. The shaded area in each of the diagrams below indicates the set determined by the definition. Such diagrams are called *Venn diagrams* for the British logician John Venn (1834–1923).

The *union* of two sets A and B is the set of elements which belongs to at least one of them:

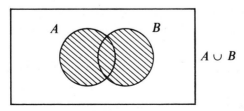

$$A \cup B = \{x \mid x \in A \text{ or } x \in B \text{ or both}\}.$$

More generally, given any collection $A_1, \ldots, A_n$ of sets, we define their union by the equation

$$\bigcup_{1 \le i \le n} A_i = \{x \mid x \in A_i \text{ for at least one } i \text{ with } 1 \le i \le n\}.$$

The *intersection* of two sets A and B is the set of elements which belong to both of them:

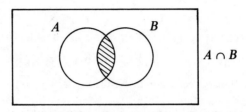

$$A \cap B = \{x \mid x \in A \text{ and } x \in B\}.$$

We say that two sets are *disjoint* if they have no common elements, $A \cap B = \varnothing$.

The *set difference*, $A - B$, of two sets A and B (in that order) is the set of those elements of A that are not in B:

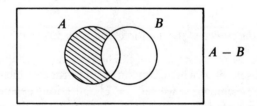

$$A - B = \{x \mid x \in A \text{ but } x \notin B\}.$$

Consider $A = \{1, 2, 3\}$ and $B = \{3, 4, 5\}$. Then $A \cup B = \{1, 2, 3, 4, 5\}$ — there is no trace of the fact that 3 was contributed by both A and B. Sometimes it is convenient to "tag" the elements of A and B so that the "history" of each element is preserved in the union. Let, then, t_A and t_B be distinct symbols, and replace A and B by the "tagged," and thus disjoint, copies

$$A \times \{t_A\} = \{(1, t_A), (2, t_A), (3, t_A)\}$$

$$B \times \{t_B\} = \{(3, t_B), (4, t_B), (5, t_B)\}.$$

Then $(A \times \{t_A\}) \cup (B \times \{t_B\}) = \{(1, t_A), (2, t_A), (3, t_A), (3, t_B), (4, t_B), (5, t_B)\}$ does indeed maintain A and B as separate.

More generally, given arbitrary sets $A_1, \ldots, A_n$ we define their *disjoint union* ΣA_i to be

$$\bigcup_{1 \le i \le n} A_i \times \{i\} = \{(x, i) \mid x \in A_i \text{ and } 1 \le i \le n\}.$$

We may also denote the union of $A_1, \ldots, A_n$ by $A_1 \cup \cdots \cup A_n$ and their disjoint union by $A_1 + \cdots + A_n$.

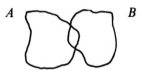

The disjoint union of A and B is the union of copies "tagged" to make them disjoint.

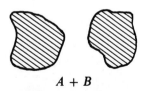

$$A + B$$

Consider, for example, **R**, the "real line" or set of real numbers. Then $\mathbf{R} \cup \mathbf{R} = \{x \mid x \in \mathbf{R} \text{ or } x \in \mathbf{R}\} = \mathbf{R}$, a single line; $\mathbf{R} + \mathbf{R} = \{(x, 1) \mid x \in \mathbf{R}\} \cup \{(x, 2) \mid x \in \mathbf{R}\}$, which is a *pair* of lines

typical point $(x, 2)$
———————————•————————————— $\mathbf{R} \times \{2\}$
———————————•————————————— $\mathbf{R} \times \{1\}$
typical point $(x, 1)$

(without disjoint unions there would be no railroads!); while $\mathbf{R} \times \mathbf{R}$ is the Cartesian plane of Figure 2.

The reader should check the following simple numerical relationships. Take care to recognize when a symbol (such as $+$) denotes a set operation.

2 Fact. *Let A and B be two finite sets, with $|A|$ and $|B|$ elements, respectively. Then*:

$$|A \times B| = |A| * |B|$$
$$|A \cup B| \le |A| + |B|$$
$$|A \cap B| \le \text{the minimum of } |A| \text{ and } |B|$$
$$|A - B| \le |A|$$
$$|A + B| = |A| + |B|. \qquad \square$$

If we are specifically interested in subsets of a fixed set S (indicated by the rectangle of the Venn diagram), we write $\bar{A}$ as an abbreviation for $S - A$:

$$\bar{A} = \{x \mid x \notin A\}$$

it being understood that the only x's we consider belong to S.

3 Fact. *For all sets A and B (as subsets of a fixed "universe" S) we have*

$$A \cap B = B \cap A$$

$$A \cup B = B \cup A$$

$$B - (A - B) = B$$

$$\bar{\bar{A}} = A$$

$$A \cap \bar{A} = \emptyset$$

$$A \cup \bar{A} = S.$$ □

Other facts can be verified as exercises.

EXERCISES FOR SECTION 1.1

1. An integer $p > 1$ is a *prime* if the only factorization of p as a product $m * n$ has $\{m, n\} = \{p, 1\}$. Thus 2, 3, and 5 are prime, but $4 = 2 * 2$ and $6 = 2 * 3$ are not. We say a number is *composite* if it is not prime.
   (a) Write down $\{x \mid x \text{ is prime and } 1 < x < 50\}$
   (b) Write down $\{x \mid x \text{ is composite and } 1 < x \leq 26\}$

2. Let $A = \{x \mid x \in \mathbf{N} \text{ and } x < 50\}$. Write down the list of elements for:
   (a) $A \cap \{x \mid x \bmod 3 = 1\}$
   (b) $A \cap \{x \mid x \operatorname{div} 7 = 2\}$
   (c) $A \cap \{x \mid x \operatorname{div} 7 = 8\}$.

3. (i) Let $A = \{\{a, b, c\}\}$. (a) What is $|A|$? (b) What is $|\mathscr{P}A|$?
   (ii) Let $B = \{a, b, c\}$. (a) What is $|B|$? (b) What is $|\mathscr{P}B|$?

4. Write down all 8 elements of $\{a, b\} \times \{a, b, c, d\}$.

5. Let $\mathbf{N}_e$ be the set of all even natural numbers and $\mathbf{N}_o$ the set of all odd natural numbers. Describe the sets
   (a) $\mathbf{N}_o \cap \mathbf{N}_e$;
   (b) $\mathbf{N}_e \cup \mathbf{N}_o$;
   (c) $\mathbf{N} \cup \mathbf{N}_e$;
   (d) $\mathbf{N} - \mathbf{N}_e$;
   (e) $\mathbf{N} \cap \mathbf{N}_e$.

6. It is true that $A \cup B = B \cup A$, and that $A \cap B = B \cap A$. We say that union ($\cup$) and intersection ($\cap$) are *commutative*. Is it true that the set difference ($-$) is commutative? Justify your answer with a careful argument.

7. Verify from the definitions that
   (a) $(A \cup B) \cup C = A \cup (B \cup C)$;
   (b) $(A \cap B) \cap C = A \cap (B \cap C)$.
   These are *associative laws*.

8. Use Venn diagrams to verify the so-called *distributive laws*:
   (a) $A \cap (B \cup C) = (A \cap B) \cup (A \cap C)$ — we say that $\cap$ distributes over $\cup$.
   (b) $A \cup (B \cap C) = (A \cup B) \cap (A \cup C)$ — we say that $\cup$ distributes over $\cap$.

(c) Point of comparison: For integers, we have $a * (b + c) = a * b + a * c$, that is, multiplication distributes over addition. Give a numerical example to show that it is *not* generally true that $a + (b * c) = (a + b) * (a + c)$.

9. Use Venn diagrams to verify (a) $\overline{A \cup B} = \bar{A} \cap \bar{B}$. Next, without using Venn diagrams, use (a) and the fact that $\bar{\bar{A}} = A$ for *any* set A to deduce that (b) $\overline{A \cap B} = \bar{A} \cup \bar{B}$. The two identities (a) and (b) are called *De Morgan's laws* after the British mathematician, logician, and author of *A Budget of Paradoxes*, Augustus De Morgan (1806–1871).

10. Find a set A with the property that A is disjoint from itself.

11. We say that a number m *divides* a number n if m divides n without a remainder, i.e., if $n \bmod m = 0$. What is the usual mathematical term for the following set

$$A = \{n \mid |D(n)| = 2\}$$

where

$$D(n) = \{m \mid m \in \mathbf{N} \text{ and } m \text{ divides } n\}$$

is the set of divisors of n?

12. Let $A_1, \ldots, A_n$ be sets with $|A_i| = m_i$ for $1 \le i \le n$. What is the cardinality of their disjoint union?

13. Prove that for any two sets A and B

$$|A \cap B| \le |A \cup B|.$$

When will it be true that two finite sets A and B satisfy $|A \cap B| = |A \cup B|$? Write down a general condition.

14. Let $|A|$ be the number of elements in the set A.
   (a) Prove by reference to a Venn diagram that for any pair of finite sets A and B, we have

$$|A \cup B| = |A| + |B| - |A \cap B|.$$

   (b) If A and B are subsets of the set S, verify that

$$|\bar{A} \cap \bar{B}| = |S| - |A| - |B| + |A \cap B|$$

   is true as a direct consequence of (a). (Hint: Use De Morgan's law.)
   (c) Complete the equation

$$|A \cup B \cup C| = |A| + |B| + |C| - \cdots + \cdots.$$

15. Give one element belonging to each of the sets given below.
   (a) $\mathbf{N} \times \mathbf{Z} \times \mathbf{R}$
   (b) $\mathscr{P}(\mathbf{N} \times \mathbf{N})$
   (c) $\mathscr{P}\varnothing \times \mathbf{N}$
   (d) $(\mathbf{N} \times \mathbf{N}) - ((\mathbf{N} - \{0, 1\}) \times \mathbf{N})$
   (e) $\mathscr{P}\mathscr{P}\mathscr{P}\varnothing$

1.2 Exponents and Series

Exponents

If we multiply n copies (with $n \geq 1$ in $\mathbf{N}$) of the number m together, we call the result m^n, which we read as m to the power n. We call n the *exponent* in this expression. We can extend this to any n in $\mathbf{Z}$ by the following rules

$$m^0 = 1$$

$$m^n = \left(\frac{1}{m}\right)^{-n} \text{ for } n < 0$$

for any non-zero number. What makes these definitions work is that they satisfy

$$m^{n_1 + n_2} = m^{n_1} * m^{n_2}$$

for any $m \neq 0$, and any integers n_1 and n_2. To see this, consider
   (i) n_1 and n_2 both greater than 0:

$$m^{n_1 + n_2} = \underbrace{m * \cdots * m}_{n_1 + n_2 \text{ times}} = \underbrace{m * \cdots * m}_{n_1 \text{ times}} * \underbrace{m * \cdots * m}_{n_2 \text{ times}} = m^{n_1} * m^{n_2}$$

   (ii) $n_2 = 0$:

$$m^{n_1 + n_2} = m^{n_1 + 0} = m^{n_1} = m^{n_1} * 1 = m^{n_1} * m^0 = m^{n_1} * m^{n_2}$$

   (iii) $n_1 > -n_2 > 0$:

$$m^{n_1 + n_2} = m^{n_1 - (-n_2)} = \frac{m^{n_1 - (-n_2)} * m^{(-n_2)}}{m^{(-n_2)}} = m^{n_1} * \left(\frac{1}{m}\right)^{(-n_2)} = m^{n_1} * m^{n_2}.$$

We leave the other cases to be checked by the reader.

We can graph the powers of m for any non-zero number m, for both positive and negative values of the exponent which in this case we write as x. The graph of Figure 4 shows the situation for an m which is greater than 1. We have "*exponential growth*" as we go positive from 0, and "*exponential decay*" as we go negative from 0.

As a result of mathematical analysis, we can fit a special smooth curve through all the points (x, m^x) for integers x, and so define m^x for all real numbers x. Note that the curve shows m^x increasing as x increases, with $m^x > 0$ for all values of x. Thus to each $y > 0$ we can associate a unique x which satisfies $y = m^x$. We call this number the *logarithm of y to the base m* and denote it by $\log_m y$.

$$x = \log_m y \quad \text{if and only if} \quad y = m^x.$$

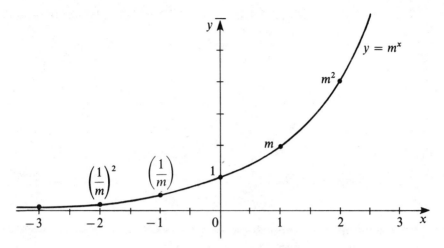

Figure 4   The curve of m^x.

The basic property of logarithms then corresponds to our fundamental law of exponents, $m^{n_1 + n_2} = m^{n_1} * m^{n_2}$, namely

$$\log_m(y_1 * y_2) = \log_m y_1 + \log_m y_2.$$

This says: to multiply two numbers we may add their logarithms, then see what number the result is the logarithm of.

SUMMING SERIES

We often encounter series of numbers built up in some regular way, and want a formula to express their sum. We show how to obtain formulas for two kinds of series — arithmetic and geometric progressions.

We say a series of numbers is an *arithmetic progression* if each term is obtained from the preceding term by *adding* the same number or *increment*. For example, the first 100 positive integers

$$1, 2, 3, 4, \ldots, 99, 100$$

form an arithmetic progression because each term is obtained from the previous term by adding 1. In general, an arithmetic progression takes the form

1 $a_1, a_2, \ldots, a_n$,   with $a_j = a + (j - 1)h$ for $1 \leq j \leq n$,

where a is the first term, h is the increment, and n is the number of terms.

Let us write S for the sum of the series in **1**. Then the sum of the terms in reverse order is also S and we have

$$S = \quad a \quad + \quad (a + h) \quad + \quad (a + 2h) \quad + \cdots + (a + (n - 2)h) + (a + (n - 1)h)$$

$$S = (a + (n - 1)h) + (a + (n - 2)h) + (a + (n - 3)h) + \cdots + \quad (a + h) \quad + \quad a$$

$$2S = (2a + (n - 1)h) + (2a + (n - 1)h) + (2a + (n - 1)h) + \cdots + (2a + (n - 1)h) + (2a + (n - 1)h)$$

Thus $2S$, double the sum, is equal to the sum of n terms, each of which is $2a + (n - 1)h$, which equals $a_1 + a_n$. Hence

$$2S = n * [a_1 + a_n]$$

2
$$S = \frac{[a_1 + a_n] * n}{2}.$$

To sum an arithmetic progression, add the first and last terms, multiply by the number of terms, and divide by 2.

We can indeed see that

$$1 + 2 = \frac{[1 + 2] * 2}{2} = 3$$

$$1 + 2 + 3 = \frac{[1 + 3] * 3}{2} = 6$$

$$1 + 2 + 3 + 4 = \frac{[1 + 4] * 4}{2} = 10$$

but it is good to have the general formula to tell us that

$$1 + 2 + 3 + \cdots + 99 + 100 = \frac{[1 + 100] * 100}{2} = 5050.$$

We say a series of numbers is a *geometric progression* if each term is obtained from the preceding term by *multiplying* by the same number or *factor*. For example, the series

$$6, 12, 24, 48, \ldots, 1536$$

forms a geometric progression because each term is obtained from the previous term by doubling. In general, a geometric progression takes the form

3
$$g_1, g_2, \ldots, g_n \quad \text{with} \quad g_j = g \cdot m^{j-1} \text{ for } 1 \le j \le n$$

where g is the first term, m is the factor, and n is the number of terms.

Let us write S for the sum of the series in **3**, and look at m times S:

$$S = g + g \cdot m + g \cdot m^2 + \cdots \qquad + g \cdot m^{n-1}$$

$$S \cdot m = \quad g \cdot m + g \cdot m^2 + g \cdot m^3 + \cdots + g \cdot m^{n-1} + g \cdot m^n.$$

Thus $S \cdot m$ differs from S by adding $g \cdot m^n$ and deleting g:

$$S \cdot m = S + g \cdot m^n - g$$

so

$$S(m - 1) = Sm - S = g(m^n - 1).$$

4
$$S = \frac{g(m^n - 1)}{(m - 1)},$$

the sum of a geometric series with first term g, factor m, and n terms.

Consider strings of *bits*, i.e., strings of 0's and 1's. There are 2 strings of length 1:

$$0, 1.$$

There are 4 strings of length 2:

$$00, 01, 10, 11.$$

There are 8 strings of length 3:

$$000, 001, 010, 011, 100, 101, 110, 111.$$

The number of strings of length ≤ 3 is

$$2 + 4 + 8 = 14$$

which agrees with formula **4** for $g = 2$, $m = 2$ and $n = 3$:

$$2\frac{(2^3 - 1)}{2 - 1} = 2 \times 7 = 14.$$

In general the number of strings of bits whose length is between 1 and n, inclusive, is $2(2^n - 1)/(2 - 1) = 2 \cdot (2^n - 1)$.

EXERCISES FOR SECTION 1.2

1. Given the digits 0, 1, 2, how many sequences (called *ternary* sequences) are there of length $\leq n$?

2. Show that $n^n = m^{n \log_m n}$.

3. For $n > 2$ which is larger, n^{2n} or $(2n)^n$?

1.3 Maps and Relations

In a phone book, each person's name is assigned a phone number

$$\text{person} \mapsto \text{phone number of person.}$$

On a football team, each player is assigned a number

$$\text{player} \mapsto \text{number on uniform.}$$

At a dance at which each person comes with a date of the opposite sex, we have an assignment of a man to each woman

woman $\mapsto$ man who is her date.

In each case we have two sets A and B and a rule, which we denote f, that assigns to each element a of A a corresponding element $f(a)$ of B.

Returning to the flow diagram of Figure 1, we see that there are four assignments:

(1) $q := 0$

(2) $r := x$

(3) $r := r - y$

(4) $q := q + 1.$

In (1), we need no values of any program variables to determine the new value of q; in (2) and (4), we need the value of one variable; while in (3), we need the values of both r and y. Given the necessary data (zero, one, or two natural numbers), we perform an appropriate mathematical operation to find a new integer and use the result as the new value of the indicated variable. The notion of map (which we will also call function) introduced in the next definition generalizes the above concepts of both mathematical operation and "real-life" assignment.

1 Definition. Given two sets A and B, a *map* or *function* f from A to B, denoted $f: A \to B$, is an assignment to each element a in A of a single element in B. We may use the notation $a \mapsto f(a)$ (note the different arrow) to indicate that $f(a)$ in B is the value assigned to a; we call $f(a)$ the *image* of a. A is called the *domain* of f, and B is the *codomain* of f. The *range* of f, denoted by:

$$f(A) = \{b \mid b \in B \text{ and } b = f(a) \text{ for some } a \text{ in } A\}$$

is the subset of B comprising the f-images of elements of A. This is the part of B that f actually "ranges over."

We can picture a map $f: A \to B$ as in Figure 5. Here we draw an arrow for each element a to its image $f(a)$ in B. Note that exactly one arrow leaves each element in A, but that zero, one, or more than one arrow may terminate on a given element of B.

The function is represented by a set of arrows from the domain to the codomain. Note that if $A = \emptyset$, then there is exactly one map $\emptyset \to B$, because the empty set of arrows is the only map possible (Figure 5b). On the other hand, if $A \neq \emptyset$, there are *no* maps of the form $A \to \emptyset$ since every element of A must go somewhere, but with $B = \emptyset$ there is no place for an arrow to go (Figure 5c).

Let us use $\mathbf{N}^2$, the set of natural numbers two-at-a-time, as shorthand for $\mathbf{N} \times \mathbf{N} = \{(m, n) \mid m \in \mathbf{N} \text{ and } n \in \mathbf{N}\}$. Let $\mathbf{N}^0$, the set of natural numbers none-at-a-time, denote a one-element set, whose single element is denoted Λ. (We shall justify this convention later; for now just notice that when we look

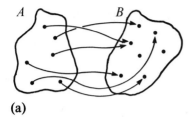

(a)

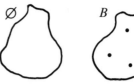

(b) This unique map $\emptyset \to B$ has no arrows.

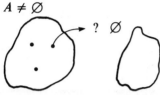

(c) There are no maps $A \to \emptyset$ for $A \neq \emptyset$.

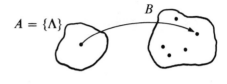

(d) Choosing an element of B specifies a map $\{\Lambda\} \to B$.

Figure 5   Maps $f: A \to B$.

at the powers of natural numbers, we take $n^0 = 1$ so that $n^{a+b} = n^a * n^b$ is true if $a = 0$.) Note that an element, b, of a set B corresponds to a map from the one-element set to B; namely, the map $f: \{\Lambda\} \to B$ with $f(\Lambda) = b$. (Figure 5d).

The mathematical operations involved in the assignments (1)–(4) can be presented in our map notation as follows:

1'. $q := 0$ can be written $q := f(\Lambda)$,   where $f: \mathbf{N}^0 \to \mathbf{N}, \Lambda \mapsto 0$
2'. $r := x$ can be written $r := g(x)$,   where $g: \mathbf{N} \to \mathbf{N}, n \mapsto n$
3'. $r := r - y$ can be written $r := h(r, y)$,   where $h: \mathbf{N}^2 \to \mathbf{N}, (m, n) \to m - n$
4'. $q := q + 1$ can be written $q := k(q)$,   where $k: \mathbf{N} \mapsto \mathbf{N}, n \mapsto n + 1$.

The observant reader will have noticed a "bug" in the map h of (3'). The difference $m - n$ only lies in $\mathbf{N}$ if $m \geq n$, and the codomain of h should be $\mathbf{Z}$, the set of all integers. This caused no problem in Figure 1, since the program tests that $r \geq y$ before making the assignment $r := r - y$. This example suggests the importance of extending the notion of a function to yield that of a *partial function* which is not necessarily defined for all elements of the domain. We will come back to the notion of partial function later.

2 Definition. We say a function $f: A \to B$ is *onto* if $f(A) = B$, i.e., every b in B is the image of some a in A.

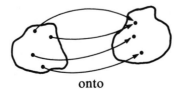

onto

We say a function $f: A \to B$ is *one-to-one* if $a \neq a'$ implies $f(a) \neq f(a')$, i.e., distinct points in A have distinct images.

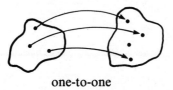

one-to-one

We say a function $f: A \to B$ is a *bijection* if it is both one-to-one and onto.

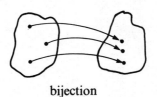

bijection

Clearly, if there is a bijection $f: A \to B$ for the finite sets A and B, then $|A| = |B|$, for if $a_1, a_2, \ldots, a_n$ is a listing of the n distinct elements which comprise the set A, then $f(a_1), f(a_2), \ldots, f(a_n)$ is a listing of all (since f is onto) the distinct (since f is one-to-one) elements in B, so that B too has n elements. Conversely, if both A and B have n elements we can write them out in any order

A	a_1	a_2	$\cdots$	a_n
B	b_1	b_2	$\cdots$	b_n

to define a bijection $f: A \to B$ by $f(a_j) = b_j$ for $1 \leq j \leq n$. We have thus proved:

3 Theorem. *Given two finite sets A and B, $|A| = |B|$ if and only if there exists a bijection $A \to B$.* $\square$

Returning to our examples, we have:

$f: \mathbf{N}^0 \to \mathbf{N}$, $\Lambda \mapsto 0$ is one-to-one but not onto;

$g: \mathbf{N} \to \mathbf{N}$, $n \mapsto n$ is a bijection;

$h: \mathbf{N}^2 \to \mathbf{Z}$, $(m, n) \mapsto m - n$ is onto, since for $r \in \mathbf{Z}$, $r \geq 0$, r is the image of $(r, 0)$; while for $r \in \mathbf{Z}$, $r < 0$, r is the image of $(0, r)$.
But $m - n = (m + 1) - (n + 1)$, and so the map is not one-to-one;

$k: \mathbf{N} \to \mathbf{N}$, $n \mapsto n + 1$ is one-to-one but is not onto — in fact, the range of the map is the set $\mathbf{N} - \{0\}$ of all positive integers.

We use the notation $A \cong B$ to indicate that A and B are *in bijective correspondence*, i.e., that there exists a bijection from A to B.

It is often important to consider the set of all maps from a set A to a set B. We use the notation $[A \rightarrow B]$ to denote this set. For example, if $A = B = \{0, 1\}$, there are four members of $[A \rightarrow B]$, as shown in the next table:

	$f(0)$	$f(1)$
f_1	0	0
f_2	0	1
f_3	1	0
f_4	1	1

At this stage, we return to the sequence $\mathbf{N}^0, \mathbf{N}^1, \mathbf{N}^2$ to discuss the general notion of *Cartesian powers*. Recall that $\mathbf{N}^3 = \{(x, y, z) | x, y, z \in \mathbf{N}\}$. Thus each element of $\mathbf{N}^3$ may be viewed as a map sending each of the three positions of the triple to $\mathbf{N}$. For example, $(13, 15, 19)$ can be thought of as a function $f: \{1, 2, 3\} \rightarrow \mathbf{N}$, with $1 \mapsto 13$, $2 \mapsto 15$, and $3 \mapsto 19$. Let us examine this point of view more closely.

Let A be any set. Consider the set $\bar{n} = \{1, 2, \ldots, n\}$ consisting of the first n positive integers. For example, $\bar{1} = \{1\}$, and $\bar{2} = \{1, 2\}$. We shall extend the notation to put $\bar{0} = \varnothing$ — fitting in with the general scheme that $\overline{n-1} = \bar{n} - \{n\}$.

Consider the set $[\bar{n} \rightarrow A]$ of all maps from $\bar{n}$ to A. There is only one map in $[\bar{0} \rightarrow A]$. We denote it by $\Lambda: \varnothing \rightarrow A$, and so we see that $[\bar{0} \rightarrow A]$ is a one-element set $A^0 = \{\Lambda\}$.

There are $|A|$ maps in $[\bar{1} \rightarrow A]$. We get a distinct map $f_a: \bar{1} \rightarrow A$ for each choice $f_a(1) = a$. We see that we can define a bijection from $A = A^1$ to $[\bar{1} \rightarrow A]$ by the rule $a \mapsto f_a$.

There are $|A|^2$ maps in $[\bar{2} \rightarrow A]$. We get a distinct map $f_{a_1, a_2}: \bar{2} \rightarrow A$ for each choice $f_{a_1, a_2}(1) = a_1$, $f_{a_1, a_2}(2) = a_2$. We see that $[\bar{2} \rightarrow A]$ is in bijective correspondence $(a_1, a_2) \mapsto f_{a_1, a_2}$ with $A^2 = A \times A$.

More generally, then, for any $n \geq 0$, we may view a function $f: \bar{n} \rightarrow A$ as the sequence of n values $(f(1), f(2), \ldots, f(n))$, so that $[\bar{n} \rightarrow A]$ is in bijective correspondence with the set of all such sequences, which set we denote by A^n. We refer to both $[\bar{n} \rightarrow A]$ and A^n as the *nth Cartesian power* of A.

Just as the powers of any number k satisfy $k^{m+n} = k^m * k^n$ for any m, n in $\mathbf{N}$, so also do we have the following.

4 Fact. *For any m, n in $\mathbf{N}$ and set A, there is a bijection*

$$A^m \times A^n \cong A^{m+n}$$

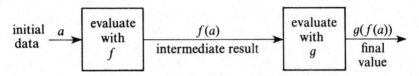

Figure 6   Sequential application of f followed by g.

which sends $((a_1, a_2, \ldots, a_m), (a'_1, a'_2, \ldots, a'_n))$ *to* $(a_1, a_2, \ldots, a_m, a'_1, a'_2,$ $\ldots, a'_n)$. *Note, in particular, that if we identify* Λ *with the empty sequence* ( ) *then we have*

$$A^m \times A^0 \cong A^m, ((a_1, \ldots, a_m), (\ )) \to (a_1, \ldots, a_m)$$

$$A^0 \times A^n \cong A^n, ((\ ), (a_1, \ldots, a_n)) \to (a_1, \ldots, a_n). \qquad \square$$

We have seen that for finite sets A and B, $A \cong B$ iff $|A| = |B|$. This implies that there is a bijection $f: A \to B$ iff there is a bijection $g: B \to A$ (Why?). We now look at this latter result in a new perspective based on function composition.

5 Definition. Given two functions $f: A \to B$ and $g: B \to C$ (where the co-domain of f equals the domain of g) we define their *composite* $g \cdot f: A \to C$ (also denoted by $A \xrightarrow{f} B \xrightarrow{g} C$) by the equality

$$g \cdot f(a) = g(f(a)) \quad \text{for each } a \text{ in } A$$

as shown in Figure 6.

It is useful to note another characterization of bijections. For any set A, the function $id_A: A \to A$, $a \mapsto a$, is called the *identity map* on A.

6 Definition. We say a map $f: A \to B$ is an *isomorphism* if it has an inverse, i.e., a map $g: B \to A$ such that

$$g \cdot f = id_A \quad \text{and} \quad f \cdot g = id_B.$$

Note that g too is an isomorphism, with inverse f.

Recall, for example, the pairing of men and women at the dance. If the pairing is exact, then the assignment $f: Men \to Women$ which maps a man to his date has as inverse the assignment $g: Women \to Men$ which maps a woman to her date. (Here *Men* is the set of men at the party; and similarly *Women*.)

7 Fact. *A map* $f: A \to B$ *is an isomorphism iff it is a bijection.*

PROOF. Suppose that f is an isomorphism. We must show that f is onto and one-to-one. Given any b in B, $b = id_B(b) = (f \cdot g)(b) = f(g(b))$ in A, so that f is onto. Now suppose that $f(a) = f(a')$. Then

$$a = id_A(a) = g \cdot f(a) = g(f(a)) = g(f(a')) = g \cdot f(a') = a'$$

so that f is one-to-one.

The other half of the proof, that each bijection is an ismorphism is left to the reader (Exercise 17).

Given the fixed "universe" S, we may associate with any subset A its *characteristic function* $\chi_A : S \to \mathscr{B} = \{T, F\}$ where

$$\chi_A(s) = \begin{cases} T & \text{if } s \in A, \\ F & \text{if } s \notin A. \end{cases}$$

Inversely, given any map $\chi : S \to \mathscr{B}$ we may associate with it the subset $\{s \mid \chi(s) = T\}$, and it is clear that each subset of S corresponds to one and only one such χ, and vice versa. In other words. we have an isomorphism between $\mathscr{P}S$, the powerset of S, and $[S \to \mathscr{B}]$, the set of all maps from S to the Boolean set $\mathscr{B}$.

8 Example. Let S be the two-element set $\{a, b\}$. Then

$$\mathscr{P}S = \{\varnothing, \{a\}, \{b\}, \{a, b\}\}$$

is a set with four elements, each of which is a subset of S. We can then tabulate the characteristic functions of these four subsets as follows:

	$\chi_\varnothing$	$\chi_{\{a\}}$	$\chi_{\{b\}}$	$\chi_{\{a,b\}}$
a	F	T	F	T
b	F	F	T	T

and we see that the subsets are thus placed in a bijective correspondence with all possible functions $\chi : S \to \{T, F\}$.

9 Fact. *Let the finite sets A and B have $|A|$ and $|B|$ elements, respectively. Then there are $|B|^{|A|}$ distinct functions from A to B. (Note the reversal — there are "$|B|$ to the $|A|$" maps from A to B.)*

PROOF. Consider the table where $m = |A|, n = |B|$

A	a_1	a_2	$\cdots$	a_m
B	$f(a_1)$	$f(a_2)$	$\cdots$	$f(a_m)$

For each of the m elements a_j of A, we may, in defining a particular $f : A \to B$, fix $f(a_j)$ to be any one of the n elements of B. We thus have a total of $(n \text{ for } a_1) * (n \text{ for } a_2) * \cdots * (n \text{ for } a_m) = n^m = |B|^{|A|}$ choices in defining the different maps from A to B. $\square$

This gives us a new proof of a result from the last section:

10 Corollary. *The finite set S has $2^{|S|}$ subsets.*

PROOF. The number of subsets of S equals the number of characteristic functions on S, namely $|\mathscr{B}|^{|S|} = 2^{|S|}$. $\square$

Note that we used a "change of viewpoint," viewing a subset as a suitable function.

PARTIAL FUNCTIONS

As mentioned earlier, partial functions are not necessarily defined for all elements of the domain. The importance of such functions in computer science may be seen by considering Figure 1 again. We have already convinced ourselves that if the input is a pair (x, y) of integers with $x \geq 0$ and $y > 0$, then the output will be a pair (r, q) with $r = x \bmod y$ and $q = x \operatorname{div} y$. We summarize this by saying that Figure 1 provides a rule for computing the function

$$f: \mathbf{N} \times (\mathbf{N} - \{0\}) \to \mathbf{N} \times \mathbf{N}, \qquad (x, y) \mapsto (x \bmod y, x \operatorname{div} y).$$

But suppose we started with $y = 0$ instead of $y > 0$. After $q := 0$, $r := x$, we have $r \geq y$, and so we enter the loop. But each time round the loop, $r := r - y$ doesn't change r, since $y = 0$. The program execution repeats the loop endlessly. We say, then, that the program of Figure 1 computes the partial function

$$f': \mathbf{N} \times \mathbf{N} \to \mathbf{N} \times \mathbf{N}$$

where

$$f'(x, y) = \begin{cases} (x \bmod y, x \operatorname{div} y), & \text{if } y > 0; \\ \text{undefined}, & \text{if } y = 0. \end{cases}$$

f' is called a partial function since on some values in the domain (namely, pairs (x, y) for which $y = 0$) there is *no* associated codomain value.

The concept of partial function is more appropriate for computer science than the traditional notion of a function. Functional relationships in computer science are usually derived from programs, and in programs infinite looping is a necessary if undesirable possibility. By allowing for undefined values, partial functions can adequately reflect the full range of functional relationships expressible by program execution.

11 Definition. Given two sets A and B, a *partial function f* from A to B, also denoted $f: A \to B$, is an assignment to each element a in a subset dom(f) of A, called the *domain of definition* of f, of a single element in B, $a \mapsto f(a)$, called the *image* of a.

The *domain of definition* of f, dom(f), is a subset of the *domain*, A, of f. The *range* of f, $f(A)$, is a subset of the *codomain*, B, of f. Since A is a subset of itself, a map or function $f: A \to B$ as defined earlier is a partial function which is *total* in the sense that dom(f) = A. (The idea of a "total partial function" may seem contradictory at first, but it makes good sense. Given a program P, we want to call the function ϕ_P that it executes a partial function if we cannot yet guarantee that it halts for all inputs, yet later investigation may allow us to guarantee termination and so conclude that ϕ_P is total.)

We now give a few more examples of partial functions and (total) functions.

The *monus* function

$$\dot{-} : \mathbf{N}^2 \to \mathbf{N}, \qquad (m, n) \mapsto \begin{cases} m - n & \text{if } m \geq n \\ 0 & \text{if } m < n \end{cases}$$

is a function (i.e., it is total), as is the function

$$\mathbf{N}^2 \to \mathbf{Z}, \qquad (m, n) \mapsto m - n.$$

Note, then, how important it is to specify the domain and codomain of a function. Thus $\mathbf{N}^2 \to B$, $(m, n) \mapsto m - n$, is total if the codomain B is $\mathbf{Z}$, but partial if $B = \mathbf{N}$.

If $\mathbf{Q}$ is the set of rational numbers (i.e., ratios m/n of integers m and n subject to the condition $n \neq 0$) we may define the partial function of division

$$\div : \mathbf{Q} \times \mathbf{Q} \to \mathbf{Q}, \qquad \left(\frac{m}{n}, \frac{p}{q}\right) \mapsto \frac{mq}{np}$$

which is *not* total since division by zero is undefined.

A change of viewpoint lets us identify a partial function with a suitable (total) function with a new codomain. Given the set B, let us form a new set $B \cup \{\perp\}$ by adding a new element $\perp$ (read "bottom," which is the T of "TOP" upside down!) which does not belong to B. Then we may view a partial function $f : A \to B$ as a total function $f^{\perp} : A \to B \cup \{\perp\}$ defined by the rule

$$f^{\perp}(a) = \begin{cases} f(a) & \text{if } a \text{ is in dom}(f) \\ \perp & \text{otherwise.} \end{cases}$$

Conversely, any function $A \to B \cup \{\perp\}$ yields a partial function.

12 Example. Let $A = \{a, a'\}$, $B = \{b\}$. Then we may tabulate the *partial* functions from A to B as

	f_1	f_2	f_3	f_4
a	b	b	—	—
a'	b	—	b	—

where — indicates that no value is defined. By changing — to $\perp$ we change each f to the corresponding $f^{\perp}$ and we see that we have a bijection from the set **Pfn**$[A, B]$ whose elements are the *partial* functions $A \to B$ to the set $[A \to B \cup \{\perp\}]$ whose elements are the *total* functions $A \to B \cup \{\perp\}$. The assignment $f \mapsto f^{\perp}$ is a bijection but we do *not* claim that any particular f_j is an isomorphism.

13 Corollary. *Given finite sets A and B, there are*

$$(|B| + 1)^{|A|} = (|B \cup \{\perp\}|)^{|A|}$$

partial *functions from A to B.*   □

We see here a powerful role of theory: to enable us to use a change of representation to convert a problem into a simpler one, or into a problem that we already know how to solve.

GRAPHS AND RELATIONS

In analytic geometry we often represent a function of the form $f: \mathbf{R} \to \mathbf{R}$ by its graph as shown in Figure 7.

We see that a typical point of the graph has coordinates $(x, f(x))$, and we may regard the graph (the curve itself, not the whole drawing which includes the axes, etc.) as a subset of $\mathbf{R} \times \mathbf{R}$. This suggests the following general definition:

14 Definition. Given any function $f: A \to B$, the *graph* of f is the subset

$$\text{graph}(f) = \{(a, b) \mid a \in A, b \in B, b = f(a)\}$$

of $A \times B$. It is clear that a subset G of $A \times B$ is the graph of some function $A \to B$ just in case for each a in A there is one and only one b in B for which $(a, b) \in G$. Moreover, given such a G, the g for which $G = \text{graph}(g)$ is defined by letting $g(a)$, for each a in A, equal the unique b in B with (a, b) in G.

For a *partial* function $f: A \to B$, only those a in the *domain of definition* of f, the subset $\text{dom}(f)$ of A, have assigned values in B. Thus for such a partial f,

$$\text{graph}(f) = \{(a, b) \mid a \in \text{dom}(f), b \in B, b = f(a)\}.$$

Then a subset G of $A \times B$ is the graph of some *partial* function $A \to B$ iff for each $a \in A$ there exists *at most one* b in B with $(a, b) \in G$. Such a G is

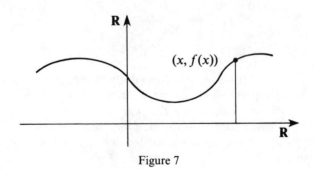

Figure 7

graph(g) for the partial function $g: A \to B$ for which $g(a)$ is the unique b with (a, b) in G, if such b exists, and for which $g(a)$ is undefined if no (a, b) is in G.

15 Example. For the partial function $h: \mathbf{N} \times \mathbf{N} \to \mathbf{N}$ which sends (m, n) to $m - n$ if $m \geq n$ but otherwise undefined,

$$\text{graph}(h) = \{(m, n, k) \mid m \geq n \text{ and } m = n + k\} \subset \mathbf{N} \times \mathbf{N} \times \mathbf{N}.$$

We have said that a (partial) function $A \to B$ can be thought of as a suitably restricted subset of $A \times B$. We next see how we may think of an arbitrary subset of $A \times B$ as a *relation*. A familiar example of a mathematical relation is *inequality*. For m and n in $\mathbf{N}$,

$$m < n \text{ if there is an integer } k > 0 \text{ such that } m + k = n.$$

We can think of this relation as a generalized function. For instead of pairing each m with a single value $f(m)$ in the way a map $f: \mathbf{N} \to \mathbf{N}$ does, it pairs each number m with a whole *set* of numbers, namely $\{n \mid m < n\}$.

Suppose, for another example, that A and B both equal the set of human beings, and that we write aRb as shorthand for "a is an aunt of b." Then we may consider the subset $\{(a, b) \mid aRb\}$ of $A \times B$, which we also denote by R. If a is a man, then $\{b \mid aRb\}$ is empty. If a is a woman with one niece and no nephews, then $\{b \mid aRb\}$ has just one element. And if the woman, a, has 17 nieces and nephews then the relationship aRb, a is the aunt of b, holds for 17 distinct elements of the set B. This motivates the general definition:

16 Definition. A relation $R: A \to B$ from A to B (note the use of the "half-arrow" to distinguish relations from functions) is a subset R of $A \times B$. We use aRb as an alternative notation for $(a, b) \in R$.

A relation may be (the graph of) a function — consider the relation $aRb =$ "b is the father of a." A relation may be (the graph of) a partial function — consider the relation $aRb =$ "b is the first-born child of a," which is partial because a person may have no children.

Just as we could pass back and forth between functions and their graphs, so may we represent relations by functions: Given a relation $R \subset A \times B$, we may associate with it a function

$$\hat{R}: A \to \mathscr{P}B.$$

(Remember, $\mathscr{P}B$ is the powerset of B, so that for each a in A, $\hat{R}(a)$ is a (possibly empty) subset of B). $\hat{R}$ sends each a in A to its set of R-relatives in B. Formally,

$$\hat{R}(a) = \{b \mid b \in B \text{ and } aRb\}.$$

Consider, for example, the relation $\leq$ on the natural numbers. We may represent it by the subset

$$R = \{(m, n) \mid m \leq n\}$$

of $\mathbf{N} \times \mathbf{N}$, or as the function $\mathbf{N} \to \mathscr{P}\mathbf{N}$ with

$$\hat{R}(m) = \{n \mid m \leq n\}.$$

In either case, we learn whether indeed $m \leq n$ for a given pair of integers — either by checking to see if $(m, n) \in R$ or by checking to see if $n \in \hat{R}(m)$.

Note that both representations correspond to the following graph where the shaded region includes the solid line $\{n \mid m \leq n\}$.

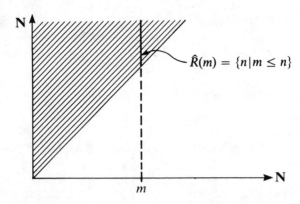

The entire shaded region represents R, while the portion of it lying above m represents $\hat{R}(m)$, just as in the graph of a *function* $f: A \to B$, the point (a, b) "lying above" a allows us to read off $f(a) = b$. More generally, an *n-ary relation* is a subset R of $A_1 \times \cdots \times A_n$ for some integer $n \geq 1$. (Thus a unary relation ($n = 1$) is just a subset of some A_1; while a binary relation ($n = 2$) is an $R: A_1 \to A_2$ in the sense of our earlier definition.)

Given a relation $R \subset A_1 \times \cdots \times A_n$, we may form its characteristic function

$$\chi_R: A_1 \times \cdots \times A_n \to \{T, F\}.$$

In fact, any function of the form $p: A_1 \times \cdots \times A_n \to \{T, F\}$ is called an *n-ary predicate*, and we see that there is a bijective correspondence between predicates and relations.

Returning to our motivating flow diagram (Figure 1 of Section 1.1), we see that a test corresponds to a relation.

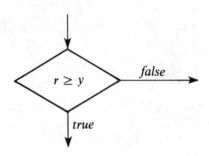

In this case, the relation is $\{(r, y) | r \geq y\} \subset \mathbf{N} \times \mathbf{N}$ and our branching convention explicitly makes use of the predicate $p: \mathbf{N} \times \mathbf{N} \to \{T, F\}$ where $p(r, y) = T$ if $r \geq y$, and $p(r, y) = F$ if not.

Finally, we look at composition of relations: If aRb means "a is the mother of b" and bSc means "b is the father of c," we may deduce that "a is the paternal grandmother of c," and call "paternal grandmother" the composite of R and S. In general, then

17 Definition. Given relations $R: A \to B$ and $S: B \to C$ their *composite* $S \cdot R: A \to C$ is defined to be the relation

$$S \cdot R = \{(a, c) | \text{there is a } b \text{ in } B \text{ with } aRb \text{ and } bSc\} \subset A \times C.$$

18 If a relation R has the same domain and codomain, $R: A \to A$, we can define the *powers* of R as follows:

$$R^0 = id_A = \{(a, a) | a \in A\}$$

$$R^1 = R$$

$$R^2 = R \cdot R$$

$$R^3 = R^2 \cdot R$$

$$\vdots$$

$$R^{n+1} = R^n \cdot R.$$

Given relations with the same domain and codomain, $S_i: A \to B$, we define their *union* $\bigcup_i S_i$ as follows: a pair (a, b) belongs to $\bigcup_i S_i$ iff $(a, b) \in S_i$ for at least one i.

19 We then write

$$R^+ \text{ for } \bigcup_{n>0} R^n$$

and

$$R^* \text{ for } \bigcup_{n \geq 0} R^n.$$

For example, if R is the relation "parent of" for human beings, then R^+ is the relation "ancestor of."

If $R: \mathbf{N} \to \mathbf{N}$ is the relation with mRn just in case $m = n + 1$, then R^+ is the relation $>$ (greater than), while R^* is the relation $\geq$ (greater than or equal to).

20 R^+ is called the *transitive closure* of R; while R^* is called the *transitive reflexive closure* of R. We shall see what these words mean when we study partial orders in Section 5.1.

EXERCISES FOR SECTION 1.3

1. Consider the relations

$$R_{\le} = \{(m, n)|m \le n\} \subset \mathbf{N} \times \mathbf{N}$$

$$R_{\ge} = \{(m, n)|m \ge n\} \subset \mathbf{N} \times \mathbf{N}.$$

Show that the intersection of $R_{\le}$ and $R_{\ge}$ is the graph of a function. Identify the function.

2. (a) Show that the following flow diagram computes z as $x * y$, the product of natural numbers x and y.
   (b) Spell out explicitly all the functions and relations used here.

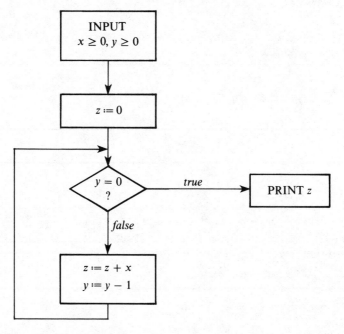

3. (a) For each of the maps below, indicate whether it is one-to-one, onto, a bijection, or none of these.

   $\mathbf{N} \to \mathbf{N}, \quad n \mapsto n^2$
   $\mathbf{Z} \to \mathbf{N}, \quad n \mapsto n^2$
   $\mathbf{N} \to \mathbf{N}, \quad n \mapsto n + 1$
   $\mathbf{Z} \to \mathbf{Z}, \quad n \mapsto n + 1$
   $\mathbf{Z} \to \mathbf{Z}_e, \quad n \mapsto 2n \quad$ where $\mathbf{Z}_e$ is the set of even integers
   $\mathbf{Z} \to \{0, 1, \ldots, m - 1\}, \quad n \mapsto n \bmod m.$

   (b) For each map which is a bijection, give the explicit definition of its inverse.

4. We say that map $g: B \to A$ is a *left inverse* (respectively, *right inverse*) of map $f: A \to B$ if $g \cdot f = id_A$ (respectively, $f \cdot g = id_B$).
   (a) Show that whenever $f: A \to B$ is one-to-one, then f has a left inverse.
   (b) Show that whenever $f: A \to B$ is onto, then f has a right inverse.

5. (i) Prove that if the relations R and S of Definition 17 are functions, then Definition 17 yields the same notion of function composition as Definition 5.

   (ii) Specialize Definition 17 to give a direct definition of the composite $g \cdot f$ of two partial functions $f: A \to B$ and $g: B \to C$.

6. Prove that if functions $f: A \to B$ and $g: B \to C$ are both one-to-one, then so is their composite.

7. Write down an explicit definition

$$R^* = \{(m, n) \mid \cdots\}$$

for the binary relation $R: \mathbf{N} \to \mathbf{N}$ where mRn just in case $|m - n| = 2$. Here $| \ |$ stands for *absolute value*,

$$|x| = \begin{cases} x & \text{if } x \geq 0 \\ -x & \text{if } x \leq 0. \end{cases}$$

The next three examples are arranged in increasing order of difficulty.

8. Let $|A| = m$ and $|B| = n$. How many isomorphisms are there from A to B? (Hint: First compare m and n. Use the factorial function in your answer.)

9. Let $|A| = m$ and $|B| = n$. How many one-to-one maps are there from A to B? (Hint: First compare m and n. Use the factorial function in your answer.)

10. Let $|A| = m$ and $|B| = n$. How many onto maps are there from A to B? (Hint: First compare m and n. Then start thinking about dividing B into m disjoint non-empty subsets.)

11. If $A = \{1, 2, \ldots, n\}$, show that a function from A to A which is one-to-one must be onto — and conversely.

12. Give an example of a function which is one-to-one but not onto from $\mathbf{N}$ to $\mathbf{N}$.

13. Show that there is an onto but not one-to-one function from $\mathbf{N}$ to $\mathbf{N}$.

14. Prove that the functions

$$f: \mathbf{N} \times \mathbf{N} \to \mathbf{N}, \qquad (m, n) \mapsto m + n$$
$$g: \mathbf{N} \times \mathbf{N} \to \mathbf{N}, \qquad (m, n) \mapsto m * n$$

are both onto but not one-to-one.

15. In the definitions to follow, f and g are binary functions on a set A, $f: A \times A \to A$ and $g: A \times A \to A$

f is said to be *commutative* if for all $a_1, a_2 \in A$

$$f(a_1, a_2) = f(a_2, a_1).$$

f is said to be *associative* if for all $a_1, a_2, a_3 \in A$

$$f(a_1, f(a_2, a_3)) = f(f(a_1, a_2), a_3)$$

f is said to be *distributive* over g if for all $a_1, a_2, a_3 \in A$

$$f(a_1, g(a_2, a_3)) = g(f(a_1, a_2), f(a_1, a_3)).$$

(i) Indicate which of the following functions $\mathbf{N} \times \mathbf{N} \to \mathbf{N}$ is commutative, or associative, or both, or neither:

$$(m, n) \mapsto m + n$$

$$(m, n) \mapsto m * n$$

$$(m, n) \mapsto m^n$$

$$(m, n) \mapsto m + (n \bmod 5)$$

$$(m, n) \mapsto lcm(m, n)$$

$$(m, n) \mapsto gcd(m, n).$$

(ii) Consider $f: \mathbf{N} \times \mathbf{N} \to \mathbf{N}$, $(m, n) \mapsto m * n$, and $g: \mathbf{N} \times \mathbf{N} \to \mathbf{N}$, $(m, n) \mapsto m + n$. Is f distributive over g? Is g distributive over f? Justify your answers.

16. Let ψ be the partial function mapping $\mathbf{N}$ to $\mathbf{N}$ defined by

$$\psi(n) = \begin{cases} n & \text{if } n \text{ is even,} \\ \text{undefined,} & \text{if } n \text{ is odd.} \end{cases}$$

Write a computer program in some higher level language such as Pascal that computes this function.

17. Complete the proof of **Fact 7**.

18. Let C equal the set of cities in the world, and for $a, b \in C$, define aRb to hold just in case there is a nonstop regularly scheduled airplane flight from a to b. What is R^0? R^1? R^2? R^+? $R*$?

CHAPTER 2

Induction, Strings, and Languages

One of the most powerful ways of proving properties of numbers, data structures, or programs is proof by induction. And one of the most powerful ways of defining programs or data structures is by an inductive definition, also referred to as a recursive definition. Section 2.1 provides a firm basis for these principles by introducing proof by induction for the set **N** of natural numbers and then looking at the recursive definition of numerical functions. Words are strings of letters and numbers are strings of digits, and in Section 2.2 we look at the set of strings over an arbitrary set, stressing how induction over the length of strings may be used for both proofs and definitions. We also introduce the algebraic notion of "semiring" which plays an important role in our study of graphs in Section 6.2.

We then turn to formal languages, subsets of the set of strings on some particular alphabet. For example, the programs of a programming language are a formal language over some fixed character set. In Section 2.3 we introduce a restricted set of formal languages, the finite-state languages, which can be associated with a class of machines called finite-state acceptors. Section 2.4 then introduces context-free grammars, a more powerful method of language definition that has proved very popular in the formal specification of the well-formed programs of such languages as Algol and Pascal. Finally, Section 2.5 looks at list processing, with special emphasis on the recursive definition and processing of s-expressions, the data structures of the programming language Lisp.

2.1 Induction on the Natural Numbers

In the course of this book a variety of proof techniques will be used to establish theorems and propositions. In this chapter we introduce one particular proof technique, *proof by induction*. Induction is especially important for computer scientists because it can serve as a principle of computation as well as a method of proof. Programs defined using *recursion* are really doing induction "backwards" during execution.

Proof by induction is a powerful method of proving that a property holds for all natural numbers. First, we check that the property holds for 0. This is called the *basis step*. Then we prove that whenever any n in $\mathbf{N}$ satisfies the property, it must follow that $n + 1$ satisfies the property. This is called the *induction step*.

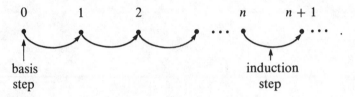

The basis step tells us that the property holds for $n = 0$. Then the induction step for $n = 0$ guarantees that the property holds for $n = 1$. Then the induction step for $n = 1$ guarantees that the property holds for $n = 2$. And so, counting up one step at a time, we see that the property must hold for all n in $\mathbf{N}$.

Proofs by induction may start at 1 as well as 0; and, indeed, proofs of this kind may start at any positive integer. A proof by induction with basis step $n = 1$ works this way: first the basis step establishes that a property holds for $n = 1$; then induction uses the $n = 1$ case to guarantee that the property holds for $n = 2$; then the induction uses the truth of the property for $n = 2$ to guarantee truth for $n = 3$, and so forth. In this case, counting up one step at a time guarantees that the property holds for all $n \in \mathbf{N}$ with $n \geq 1$.

We give two examples of proof by induction.

In Figure 8, we see how successively larger squares are formed by adding borders. For example, the shaded border around the 2×2 square in Figure 8 contains $1 + 2 * 2 = 5$ squares. We observe that

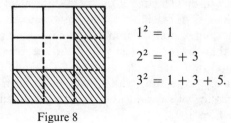

$$1^2 = 1$$
$$2^2 = 1 + 3$$
$$3^2 = 1 + 3 + 5.$$

Figure 8

This suggests the hypothesis that, for every $n \geq 1$, we have that n^2 equals the sum of the first n odd numbers.

$$n^2 = 1 + 3 + \cdots + (2n - 1).$$

We prove this by induction.

Basis Step: (Since we only want to prove our result for $n \geq 1$, we take $n = 1$ as our basis step.) For $n = 1$, $1^2 = 1$, and the property holds.

Induction Step: Suppose the hypothesis holds for n, so that

$$n^2 = 1 + 3 + \cdots + (2n - 1). \tag{1}$$

Then the sum of the first $n + 1$ odd numbers equals

$$1 + 3 + \cdots + (2n - 1) + (2n + 1) = n^2 + (2n + 1) = (n + 1)^2 \quad \text{by (1)}$$

and so our hypothesis holds for $n + 1$.   □

Many students on first reading a proof by induction are stymied by the assumption: "Suppose the hypothesis holds for n." "What," they ask, "if it *doesn't* hold for n?" Perhaps it helps to expand the statement as follows:

We use n as an algebraic variable which can be made to take any value in the set $\mathbf{N} = \{0, 1, 2, 3, \ldots\}$ of natural numbers. At any time we assume that n has a single fixed value in $\mathbf{N}$. Returning to our hypothesis, we do not yet know whether or not it holds for all n. That is what we are trying to prove. But suppose that we now assign to n a value for which the hypothesis *does* hold. We now know that if we add one to this particular value — we denote the result by $n + 1$ — then the hypothesis also holds for this value.

Let us, then, use $P(n)$ as our shorthand for "the hypothesis holds for n." If we want to prove that $P(n)$ for all $n \geq 0$, the basis step requires us to prove $P(0)$. The induction step requires us to prove that $P(n)$ implies $P(n + 1)$, i.e., for every value of n in $\mathbf{N}$ for which $P(n)$ is true we *must* have that $P(n + 1)$ is true. Note that the induction step tells us *nothing* about $P(n + 1)$ if $P(n)$ is *not* true.

If we can verify *both* the basis step *and* the induction step, we can infer the truth of $P(n)$ for all n. For example, we can see that $P(3)$ holds because

$P(0)$ is true by the basis step.
$P(0)$ implies $P(1)$ by the induction step with $n = 0$.
Thus $P(1)$ is true.
$P(1)$ implies $P(2)$ by the induction step with $n = 1$.
Thus $P(2)$ is true.
$P(2)$ implies $P(3)$ by the induction step with $n = 2$.
Thus $P(3)$ is true.

Now it was rather boring to write this explicit proof out for $n = 3$. Imagine how tedious it would be to write the proof out for $n = 1,000,000$ — it would have 2,000,001 lines! And that's the whole point of proof by induction. We *don't* have to write out a separate proof for every natural number we are interested in. Just a proof of the basis ($P(0)$ is true) and the induction step (for every natural number n, it is the case that the truth of $P(n)$ guarantees the truth of $P(n + 1)$) will do.

1 Example. Consider the claim "$n > 7$." The induction step is clearly correct, since *if* $n > 7$ then certainly $n + 1 > 7$. But in this case the basis step *fails*, because it is not true that $0 > 7$. So we *cannot* prove by induction that for every natural number n, it is true that $n > 7$. (We can, however, prove by induction that "$n > 7$" is true for every $n \geq 8$ if we choose $n = 8$ as our basis step.)

2 Example. Consider the claim "$n < 7$." Here the basis step works, $0 < 7$, but the induction step fails — we cannot prove "$n < 7$ implies $n + 1 < 7$" because it fails for the particular value $n = 6$.

We now give a new proof of Fact 1 in Section 1.1:

3 Theorem. *Each n element set has 2^n subsets.*

PROOF BY INDUCTION

Basis Step: For $n = 0$, the only n element set is the empty set $\emptyset$ which has only one subset, namely $\emptyset$. But $2^0 = 1$, establishing the basis.

Induction Step: Suppose that every n element set has 2^n subsets. We must show that this guarantees that $|A| = n + 1$ implies $|\mathscr{P}A| = 2^{n+1}$. Let then $A = B \cup \{a\}$, where B has n elements, and a is an element not in B. Each subset of A either does or does not contain a.

$$\mathscr{P}A = \{S \,|\, S \subset A, a \in S\} \cup \{S \,|\, S \subset A, a \notin S\}$$

$$= \{T \cup \{a\} \,|\, T \subset B\} \cup \{T \,|\, T \subset B\}.$$

Thus $|\mathscr{P}A| = |\mathscr{P}B| + |\mathscr{P}B| = 2^n + 2^n$, by hypothesis on n, $= 2^{n+1}$.   □

Before we go further, it will be useful to distinguish a number from the notation which represents it. We may write

$\mathbf{N} = \{0, 1, 2, 3, 4, 5, \ldots\}$
or $\mathbf{N} = \{\text{zero, one, two, three, four, five}, \ldots\}$
or $\mathbf{N} = \{0, \text{I, II, III, IV, V}, \ldots\}$
or $\mathbf{N} = \{0, 1, 10, 11, 100, 101, \ldots\}$

and we realize that these are notational variants of each other. In each case, we first write down an element (0 or zero, say) which represents the number we use to count an *empty* pile of stones; and when we have written a representation of the number n we use to count the stones in a given pile, we may then write down the representation of the number $\sigma(n)$ — the *successor* of n — that counts the stones in the pile obtained by adding a single stone to the previous pile. The Italian mathematician Giuseppe Peano (1858–1932) summarized our intuitions about the non-negative integers in the following:

4 Peano's Axioms. The set **N** contains an element 0, and is equipped with a function which assigns a number $\sigma(n)$ to each number n in such a way that the following properties hold:

P1. For no n in **N** does $\sigma(n) = 0$. (0 is the first member of **N** and so is the successor of no other number.)

P2. If $m \neq n$ then $\sigma(m) \neq \sigma(n)$. (Distinct numbers have different successors.)

P3. If a subset S of **N** contains 0 and contains $\sigma(n)$ for each n in S, then in fact S is precisely **N**.

Note that P3 states the validity of proof by induction. For let $P(n)$ be a statement that is either true or false for each natural number n, and let S be the set of n for which $P(n)$ is true. (In other words, $P = \chi_S: \mathbf{N} \to \{T, F\}$, the characteristic function of S as a subset of **N**.) Then P3 says precisely: If $P(0)$ is true (basis step), and if $P(n)$ is true implies $P(\sigma(n))$ is true (induction step), then $P(n)$ is true for every n in **N**.

RECURSION

So far, we have talked about verifying the properties of natural numbers by induction. But we can also use induction to *define* new constructs. A mathematician will call this type of definition an inductive definition, but a computer scientist will call it a *recursive definition*. We shall discuss why after looking at the "standard example."

The *factorial function* $\mathbf{N} \to \mathbf{N}$, $n \mapsto n!$, assigns to 0 the value 1, and assigns to each $n > 0$ the value $n! = n * (n - 1) * \cdots * 2 * 1$, i.e., the product of the first n positive integers:

$$0! = 1$$

$$1! = 1 = 1 * 0!$$

$$2! = 2 * 1 = 2 = 2 * 1!$$

$$3! = 3 * 2 * 1 = 6 = 3 * 2!$$

$$4! = 4 * 3 * 2 * 1 = 24 = 4 * 3!$$

Thus, for each natural number n either $n = 0$ and $n! = 1$, or $n > 0$ and $n! = n * (n - 1)!$. Going further, this permits us to define $n!$ by the following scheme:

5 Recursive Definition of the Factorial Function

Basis Step: If $n = 0$, then $n! = 1$.

Recursion Step: If $n!$ is already defined, then we can define $(n + 1)!$ to be $(n + 1) * n!$.

We rename the induction step of our definition "the recursion step" to indicate that the function we are defining *recurs* in the actual definition (i.e., on the right-hand side of the equation — the definition — not just on the left-hand side, which simply tells us the notation for what is being defined).

This recursive definition of a function $f: \mathbf{N} \to \mathbf{N}$, then, acts just like an inductive proof. The basis step tells us how to define $f(0)$ explicitly. The recursion step tells us that *if* we already know how to define $f(n)$ (for some particular value n in $\mathbf{N}$) *then* we can define $f(n + 1)$ for the successor of that value of n.

Such a definition starts with 0 and works up one step at a time to define $f(n)$ for larger and larger natural numbers. But if we want to compute $f(n)$ we work the other way around. Consider

$$4! = 4 * 3! \tag{2}$$
$$= 4 * 3 * 2!$$
$$= 4 * 3 * 2 * 1!$$
$$= 4 * 3 * 2 * 1 * 0!$$
$$= 4 * 3 * 2 * 1 * 1$$
$$= 24.$$

The idea is this. To compute $f(n)$ we check to see if $n = 0$, in which case we apply the basis step of the definition. Otherwise we subtract 1 from n to *reduce the problem* to that for a smaller integer. And so it goes through n *reduction steps* (running the recursion step backwards, as it were) until we return to the basis and complete the computation.

Many programming languages allow us to use a recursive definition as part of a program:

$$n! := \text{if } n = 0 \textbf{ then } 1 \textbf{ else } n * (n - 1)! \tag{3}$$

tells the computer

test to see if $n = 0$;

if the answer is YES, then the value of $n!$ is 1;

if the answer is NO, then the value of $n!$ is $n * (n - 1)!$

and the computer must call the instruction (3) again (but with n replaced by $(n - 1)$) to compute $(n - 1)!$. And so on, just as in (2) for $n = 4$.

The interesting thing about the program (3) is that it cannot be represented as a flow diagram. However, we can draw a flow diagram as in Figure 9 for a different way of computing $n!$. Here we have a loop with a variable m to count how many times around the loop remain (we count down from n to 0) and a variable w which, after j times around the loop, will

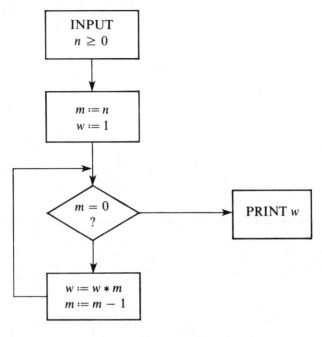

Figure 9

hold the value $n * (n - 1) * \cdots * (n - j + 1)$ — which for $n = j$ is just $n!$, the desired result.

Note that this program is *not* recursive. Nowhere in this program for factorial does factorial itself get mentioned, as it did in the recursive program (3). Instead the solution is by *iteration*: we build a loop and *iterate* (repeat) it again and again until we have satisfied the test that allows the computation to exit from the loop.

NUMBERS AS STRINGS OF DIGITS

We know what natural numbers are, and we know what the decimal representations of these numbers are — they are strings of digits such as 0 or 3 or 13 or 99 or 659432. We also "know" how to add 1 to any number in decimal notation — in the above cases we get 1, 4, 14, 100, and 659433, respectively. But in what sense do we know? Most students find it very hard to write down a precise algebraic definition for this "decimal successor" function. (Try it for five minutes before you read on.) Our aim is to show how a *recursive definition* can be used to solve the problem.

Consider again Peano's axioms, introduced in this section. These axioms say we can construct all of **N** via a *basis step*, which gives us 0, and repeated application of the *induction step*, which gives us $\sigma(n)$ once we have n. Recall that $\sigma(n) = n + 1$, the successor of n. More generally, an inductive or recursive definition of a set (generalizing from $\mathbf{N}$) is one which defines certain basic elements, and then defines a process which can be applied repeatedly to give more and more elements.

6 Inductive Definition of N_D, the Set of Decimal Representations of the Natural Numbers. Let $D = \{0, 1, 2, 3, 4, 5, 6, 7, 8, 9\}$ be the set of decimal digits.

(i)   *Basis Step*: Each d in D belongs to $\mathbf{N}_D$.
(ii) *Induction Step*: If w is in $\mathbf{N}_D$ and d is in D, then wd (the set of digits obtained by following w by the digit d) is also in $\mathbf{N}_D$.
(iii) Nothing belongs to $\mathbf{N}_D$ save by (i) and (ii).

For example, to see that 1357 is in $\mathbf{N}_D$, we note that 1 is in $\mathbf{N}_D$ by (i), so that 13 is in $\mathbf{N}_D$ by (ii), so that 135 is in $\mathbf{N}_D$ by (ii), and so 1357 is in $\mathbf{N}_D$ by (ii). (But note Exercise 11.)

Before going further, we need the table that tells us in which order the digits occur. We define $m: D \to D$ by the display

d	0	1	2	3	4	5	6	7	8	9
$m(d)$	1	2	3	4	5	6	7	8	9	0

This is an exhaustive, explicit definition. All possible cases are in the table. By contrast, when we deal with large or infinite sets, we need a rule rather than an explicit table to tell us the definition of a function.

Now we have to define $\sigma: \mathbf{N}_D \to \mathbf{N}_D$ so that for any string w of digits, $\sigma(w)$ represents the successor of the number given by w. Clearly, for each digit, $\sigma(d) = m(d)$, except when $d = 9$, and then $\sigma(9) = 10$.

Now what about a multidigit number? If it does not end in a 9, we simply replace the last digit d by $m(d)$. If d is 9, we must replace it by 0, and "carry 1."

7 Definition. The successor function $\sigma: \mathbf{N}_D \to \mathbf{N}_D$ for natural numbers in decimal notation is defined recursively:

Basis Step: If $v = d$ in D, then

(a)   If $d \neq 9$, $\sigma(v) = m(d)$
(b)   If $d = 9$, $\sigma(v) = 10$.
      Recursion Step: If $v = wd$ with w in $\mathbf{N}_D$, d in D, then
(a)   If $d \neq 9$, $\sigma(v) = wm(d)$
(b)   If $d = 9$, $\sigma(v) = \sigma(w)0$.

8 Example

$$\sigma(999) = \sigma(99)0 = \sigma(9)00 = 1000$$

$$\sigma(1399) = \sigma(139)0 = \sigma(13)00 = 1400$$

$$\sigma(236) = 23m(6) = 237.$$

In the last section, we studied induction on **N**. {0} was the basis, and the passage from n to $n + 1$ was the induction step. Here we consider $\mathbf{N}_D$ with the set D of digits as the basis. The passage from any string w to each of the strings wd, one for each digit in D, is the induction step.

For another look at the inductive treatment of numbers in decimal notation, we present addition in terms of a recursive definition of $\mathbf{N} \times \mathbf{N} \to \mathbf{N}$, $(m, n) \mapsto m + n$ based on the intuition "$m + n$ is the number of stones in a pile of stones obtained by merging a pile of m stones with a pile of n stones."

9 Recursive Definition of Addition. Having already defined the successor function, we may define $m + n$ by induction on n as follows:

Basis Step: $m + 0 = m$.
Induction Step: Given that $m + n$ is already defined, we set

$$m + \sigma(n) = \sigma(m + n).$$

10 Example

$$
\begin{aligned}
4 + 3 &= 4 + \sigma(2) && \text{since } 3 = \sigma(2) \\
&= \sigma(4 + 2) && \text{by induction step} \\
&= \sigma(4 + \sigma(1)) && \text{since } 2 = \sigma(1) \\
&= \sigma(\sigma(4 + 1)) && \text{by induction step} \\
&= \sigma(\sigma(4 + \sigma(0))) && \text{since } 1 = \sigma(0) \\
&= \sigma(\sigma(\sigma(4 + 0))) && \text{by induction step} \\
&= \sigma(\sigma(\sigma(4))) && \text{by basis step} \\
&= \sigma(\sigma(5)) && \text{since } \sigma(4) = 5 \\
&= \sigma(6) && \text{since } \sigma(5) = 6. \\
&= 7 && \text{since } \sigma(6) = 7.
\end{aligned}
$$

Of course, this is a tedious way to add numbers, and neither humans nor computers usually add numbers this way. Moral: The first program you find to do a job may not be the best one.

EXERCISES FOR SECTION 2.1

1. Prove by induction that

(a)
$$1^2 + 2^2 + 3^2 + \cdots + n^2 = \frac{n(n + 1)(2n + 1)}{6}$$

(b)
$$\frac{1}{1 \cdot 2} + \frac{1}{2 \cdot 3} + \cdots + \frac{1}{n(n + 1)} = \frac{n}{n + 1}$$

(c)
$$\frac{1}{2} + \frac{2}{2^2} + \frac{3}{2^3} + \cdots + \frac{n}{2^n} = 2 - \frac{n + 2}{2^n}.$$

2. Prove the following inequalities by induction:

(a) $(1 + \frac{1}{2})^n \geq 1 + \frac{1}{2}n$.

(b) For $x > -1$, $(1 + nx)^n \geq 1 + nx$.

3. Let $P(n)$ be the assertion "In any group of n people, all the people have the same color hair." What is wrong with the following proof that $P(n)$ is true for all $n \geq 1$?

Basis Step: $P(1)$ is certainly true.

Induction Step: Suppose $P(n)$ is true. Given a group of $n + 1$ people, arrange them in order:

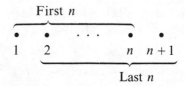

Since $P(n)$ is true, the first n have the same color hair, and the last n all have the same color hair. But the nth person belongs to both groups, and so all $n + 1$ people must have the same color hair. Thus $P(n + 1)$ is true.

4. The recursive definition of addition also works for numbers in binary notation. To check this, compute:

(i) $1101 + 101$

(ii) $1010 + 11$.

5. Assuming that the function $+ : \mathbf{N} \times \mathbf{N} \to \mathbf{N}$ of addition is already available, we can give a recursive definition of multiplication as follows:

Basis Step: $m * 0 = 0$

Recursion Step: $m * \sigma(n) = m * n + m$.

(i)   Use this definition to compute $6 * 4$. (Just add all necessary pairs of numbers directly — don't use the recursive definition **9**.)

(ii) Write a flow chart for an *iterative* program which multiplies two numbers, using addition as a primitive operation.

6. Write a recursive definition of the exponential, $\exp : \mathbf{N} \times \mathbf{N} \to \mathbf{N}$, $(m, n) \mapsto m^n$, making use of multiplication as a primitive operation.

7. Write a recursive definition of the monus operation,

$$m \div n = \begin{cases} m - n & \text{if } m > n \\ 0 & \text{otherwise} \end{cases}$$

using only the successor function as a primitive operation. (Hint: First define by recursion a predecessor function, $p(n) = n \div 1$.)

8. Prove by induction that $2^n < n!$, for all $n \geq 4$.

9. Prove by induction that $n^3 + 2n$ is divisible by 3, for all nonnegative integers n.

10. Give an inductive definition of the set $\{10, 15, 20, \ldots\}$, i.e., all multiples of 5 starting with 10, using only addition (of two numbers) in the induction step.

11. Modify definition **6** to avoid generating strings like 0116, without losing 0.

12. Give an inductive definition of the set $\{n \mid n$ is a non-negative multiple of 3 and 5$\}$, using only addition in the induction step.

2.2 The Strings over an Arbitrary Set

The notion of a string of symbols is familiar to every computer scientist. In this section we consider the mathematical properties of this concept. To do this we will fix an abstract alphabet or set of characters X, and we will consider the set of all finite strings of symbols from X. This set of strings is called X^*. One particular string, the *empty string* (it has length 0 and is denoted by Λ), plays a special role in the study of X^*, in much the same way that 0 plays a special role for $\mathbf{N}$. We now formally define X^*.

1 Definition. For any set X, X^* is the set of all finite strings (or sequences) over the alphabet X. We write a typical element of X^* as $w = (x_1, \ldots, x_n)$ or $x_1 \cdots x_n$, where $x_i \in X$ for every $1 \leq i \leq n$, and say that its *length* $\ell(w)$ is n. We also include in X^* the *empty string* $\Lambda = (\ \ )$, with $\ell(w) = 0$.

The set X^+ is the set $X^* - \{\Lambda\}$ obtained from X^* by deleting the empty string.

Given any two strings $w = (x_1, \ldots, x_m)$ and $w' = (x'_1, \ldots, x'_n)$ we may *concatenate* them to obtain

$$ww' = (x_1, \ldots, x_m, x'_1, \ldots, x'_n).$$

We note that

    (a) $\ell(ww') = \ell(w) + \ell(w')$ — the length is like a logarithm!

    (b) $w(w'w'') = (ww')w''$ and so we may write either as $ww'w''$.

    (c) $w\Lambda = \Lambda w = w$.

2 Example. The set $\mathbf{N}_D$ of decimal representations of numbers **(2.1.6)** equals D^+.

3 Example. If $X = \emptyset$, the empty set, X^* has only one element, namely Λ: $\emptyset^* = \{\Lambda\}$. But for every nonempty set X, we see that X^* is an infinite set. For example, $\{1\}^* = \{\Lambda, 1, 11, 111, \ldots, 1^n, \ldots\}$ where a typical element is a string

$$\underbrace{11 \cdots 1}_{n \text{ times}}$$

of n 1's which we denote 1^n. Thus we have a bijection $\mathbf{N} \cong \{1\}^*$, $n \mapsto 1^n$. Caution: 1^n is *string* notation (n 1's in a row) not *number* notation (n 1's multiplied together).

Remember that, in a sequence, the length of the sequence and the position of each element all matter, so that $m \neq n$ implies $1^m \neq 1^n$ in string notation. On the other hand, a set is defined by its elements, irrespective of their order; e.g., $\{1, 1, 1\} = \{1\}$ and the set $\mathscr{P}\{1\}$ is finite: $\mathscr{P}\{1\} = \{\varnothing, \{1\}\} = \{\{1\}, \varnothing\}$.

In Section 1.1 we introduced the notion of *disjoint union* of sets while in Section 1.3 we introduced *Cartesian powers*. We may combine these notions to give an alternative definition of (a set in bijective correspondence with) X^*:

$$X^* = \sum_{n \geq 0} X^n = X^0 + X^1 + X^2 + \cdots + X^n + \cdots$$

$$= \{\Lambda\} + X + X \times X + \cdots + \underbrace{X \times \cdots \times X}_{n \text{ times}} + \cdots.$$

This expression says that X^* is the union of the length 0 strings, the length 1 strings, the length 2 strings, etc. A typical element of X^n is $(x_1, x_2, \ldots, x_n)$ for that same n. Remember that the different terms are *tagged* to make them disjoint in a disjoint union:

$$\sum_{n \geq 0} X^n = \bigcup_{n \geq 0} X^n \times \{n\}.$$

Thus, to inject $(x_1, x_2, \ldots, x_n)$ into the disjoint union $X^0 + X^1 + X^2 + \cdots + X^n + \cdots$ we tag it with the *index n*. Thus in this presentation of X^*, the typical element is written

$$((x_1, x_2, \ldots, x_n), n)$$

which explicitly records the length n along the sequence itself. Concatenation is then $((x_1, \ldots, x_m), m) \cdot ((x'_1, \ldots, x'_n), n) = ((x_1, \ldots, x_m, x'_1, \ldots, x'_n), m + n)$. There is an obvious bijective map from X^* to the disjoint union, sending $(x_1, \ldots, x_n)$ to $((x_1, \ldots, x_n), n)$.

As an exercise in inductive definition, we turn to a third definition of X^*, giving yet a third way of denoting a list of n elements of X. We formalize, for any set X, the formation of longer strings by adding elements x of X at the left-hand end. (This is very similar to Definition **2.1.6**, but there we formed longer and longer elements of $\mathbf{N}_D$ by adding new digits to the right-hand end of a decimal string.)

4 Definition. For any set X, we simultaneously define X^* and the *length function* $\ell : X^* \to \mathbf{N}$ inductively by

Basis Step: X^* contains a distinguished element, written Λ, which we call the empty string. We set $\ell(\Lambda) = 0$.

Induction Step: If w is an element of X^* and x is an element of X, then (x, w) is an element of X^*. We may abbreviate (x, w) as xw. We set $\ell(xw) = \ell(w) + 1$.

(We sometimes use (w, x) in the induction step to define X^*.)

We then define a map conc: $X^* \times X^* \to X^*$, $(w, w') \mapsto ww'$, called *concatenation*, by induction on the length of w as follows:

Basis Step: conc$(\Lambda, w') = w'$
Induction Step: conc$(xv, w') = (x, \text{conc}(v, w'))$ where $v \in X^*$.

As an exercise, the reader should verify that properties (a), (b), and (c), mentioned after **1** follow from the inductive definition of **4** above.

Note that our inductive definition offers the notation $(x_1, (x_2, (x_3)))$ for what we normally write as (x_1, x_2, x_3) or $x_1 x_2 x_3$. The reader familiar with Lisp will recognize Λ as being another notation for what Lispers call NIL. Lisp is a list-processing language especially important in Artificial Intelligence (AI). (AI is a field of computer science which studies computational processes involved in problem-solving, natural language understanding, etc. A very readable introduction to AI for the nonspecialist is Margaret Boden's "Artificial Intelligence and Natural Man," Basic Books, 1977.) In Section 2.5, we will build on our knowledge of X^* to define the basic data stucture of Lisp.

X^*, the set of finite strings over some finite alphabet X, is the fundamental object of *formal language theory*. We define a *language over X* to be any subset of X^*. Thus, the set of words of English is a language over the alphabet $X = \{a, b, \ldots, y, z\}$. Likewise $L = \{a^n b^n | n \geq 1\}$, consisting of a string of a's followed by a string of b's of equal length, is a language over $X = \{a, b\}$.

Three basic operations are important in language theory.

5 Definition. Let X be a fixed finite alphabet, and let A and B be languages over X. Then

1. The *union* of A and B, $A \cup B = \{w | w \in A \text{ or } w \in B\}$. $A \cup B$ is sometimes written as $A + B$ in language theory, where $+$ does *not* denote the disjoint union.
2. The *concatenation* $A \cdot B = \{w_1 w_2 | w_1 \in A \text{ and } w_2 \in B\}$ consists of all strings from X^* that are made up of a string from A followed by a string from B.
3. The *iterate* (or *Kleene star*) $A^* = \{w_1 \cdots w_n | n \geq 0, \text{ each } w_i \in A\}$ is made up of all strings from X^* that can be obtained by concatenating any finite number of strings from A. (Notice that, as with X^*, $\Lambda \in A^*$, corresponding to the case where $n = 0$).

6 Examples. Suppose we take $X = \{a, b\}$, and set $A = \{a\}^*$, $B = \{b\}^*$. Then A contains all finite strings of a's, B contains all finite strings of b's, and $C = A \cup B$ contains any finite strings of a's as well as any finite string of b's. Strings of $A \cdot B$ are made up of a finite string of a's followed by any finite string of b's. $A \cdot B$ includes *aabbb*, *aaa* (there may be no b's at all), and *bbbb* (there may be no a's at all). The strings *ba* and *abba* are not in $A \cdot B$, since no b may precede any a.

If A is any language, it is often useful to view A^* as the infinite (disjoint) union of successively longer concatenations of A with itself:

$$A^* = A^0 + A + A \cdot A + A \cdot A \cdot A + \cdots + \underbrace{A \cdot \ldots \cdot A}_{n \text{ times}} \cdots .$$

Here A^0 represents $\{\Lambda\}$, and a typical element of this sum,

$$\underbrace{A \cdot \ldots \cdot A}_{n \text{ times}}$$

is made up of all possible concatenations of any n strings from A.

7 Example. The language $(\{a\} \cup \{bb\})^* = \{\Lambda\} \cup (\{a\} \cup \{bb\}) \cup (\{a\} \cup \{bb\})$ $\cdot (\{a\} \cup \{bb\}) \cdots = \{\Lambda\} \cup \{a\} \cup \{bb\} \cup \{aa\} \cup \{abb\} \cup \{bba\} \cup \{bbbb\} \cdots =$ all strings of a's and b's in which b's occur consecutively as doubletons.

8 Example. The language $(\{a^4\} \cup \{a^6\})^* = \{\Lambda\} + \{a^4\} + \{a^6\} + \{a^8\} + \{a^{10}\} + \cdots + \{a^{2n}\} + \cdots = \{a^{2k} | k \geq 2\}$.

MONOIDS, GROUPS, AND SEMIRINGS

In the remainder of this section, we show how the above operations on X^* can be placed in an algebraic perspective. The material here will not be used until Section 6.2 and so may be omitted at a first reading. We write the triple $(X^*, \text{conc}, \Lambda)$ to mean that we are to consider as an object of study the set X^* together with the operator conc: $X^* \times X^* \to X^*$ and the distinguished element $\Lambda \in X^*$. Our $(X^*, \text{conc}, \Lambda)$ triple has two properties that are of general interest in computer science and mathematics. First of all conc is *associative*: $(w_1 w_2)w_3 = w_1(w_2 w_3)$ — that is, $\text{conc}(\text{conc}(w_1, w_2), w_3) = \text{conc}(w_1, \text{conc}(w_2, w_3))$. And secondly the empty string is an *identity for conc*: $\text{conc}(\Lambda, w) = \text{conc}(w, \Lambda) = w$.

Whenever a triple of this kind has these two properties, then it is an example of structure called a *monoid*.

9 Definition. A *monoid* is a triple (M, m, e) where M is a set, $m: M \times M \to M$ is a binary operation on M, and e is an element of M, subject to the conditions:

1. m is *associative*: For all elements x, y, z of M we have

$$m(m(x, y), z) = m(x, m(y, z)).$$

2. e is an *identity* for m: For all elements x of M we have

$$m(x, e) = m(e, x) = x.$$

10 Example. $(\mathbf{N}, +, 0)$ is a monoid:

$$(m + n) + p = m + (n + p)$$

$$m + 0 = 0 + m = m.$$

$(\mathbf{N}, *, 1)$ is a monoid:

$$(m * n) * p = m * (n * p)$$

$$m * 1 = 1 * m = m.$$

We say that 0 is the *additive identity* for $\mathbf{N}$, and that 1 is the *multiplicative identity* of $\mathbf{N}$.

Why do mathematicians introduce abstract concepts like "monoid"? Because it is often possible to prove a property once and for all in the general setting, and then use it in any special case without any further work. Here is a simple example:

11 Fact. *A monoid has only one identity.*

PROOF. Suppose that e and e' are both identities for m. Then

$$m(e, e') = e \text{ because } e' \text{ is an identity;}$$

$$m(e, e') = e' \text{ because } e \text{ is an identity;}$$

and hence $e = e'$. $\qquad\qquad\qquad\qquad\qquad\qquad\qquad\qquad\qquad\qquad$ □

This saves us an individual investigation in each case to verify the uniqueness of Λ and 0 as the identities of $(X^*, \text{conc}, \Lambda)$ and $(\mathbf{N}, +, 0)$, respectively.

There is an important difference between the set of all integers and the set of all natural numbers. Each integer n in $\mathbf{Z}$ has an *additive inverse* $-n$, that is

$$n + (-n) = 0 = (-n) + n.$$

By contrast, the only n in $\mathbf{N}$ with an additive inverse is 0,

$$0 + 0 = 0,$$

since the minus of any positive integer is negative. We have another general definition:

12 Definition. Let $(M, \cdot, e)$ be a monoid. Then we say that element x is an *inverse* for the element y if

$$x \cdot y = e = y \cdot x.$$

A monoid is called a *group* if every element of M has an inverse.

We now look at Example 10 in more detail.

1. $(\mathbf{N}, +, 0)$ is a *commutative* monoid, i.e., not only is it a monoid, but, moreover, $m + n = n + m$ for all m, n in $\mathbf{N}$.
2. $(\mathbf{N}, *, 1)$ is a monoid. (It is commutative, too, but we shall not use this fact here.)
3. Multiplication distributes over addition, i.e.,

$$m * (n + p) = m * n + m * p$$

for all m, n, and p in $\mathbf{N}$. This calls for yet another general definition:

13 Definition. A quintuple $(S, +, \cdot, 0, 1)$ is called a *semiring* with additive identity 0 and multiplicative identity 1 if

1. $(S, +, 0)$ is a commutative monoid;
2. $(S, \cdot, 1)$ is a monoid;
3. $\cdot$ distributes over $+$:

$$a \cdot (b + c) = a \cdot b + a \cdot c \quad \text{for all } a, b, c \text{ in } S.$$

One reason for the interest of this concept to computer scientists is given by the following example. Another is given in Exercise 6. We shall return to semirings when we study connection matrices for graphs in Section 6.2.

14 Observation. *Let $L_X = \mathscr{P}X^*$ be the set of all languages over the alphabet X. Then union, $\cup$, has an identity $\varnothing$, the empty subset:*

$$A \cup \varnothing = A = \varnothing \cup A \quad \text{for all } A \subset X^*.$$

Moreover, $(L_X, \cup, \varnothing)$ is a commutative monoid.

Concatenation, $\cdot$, has an identity $\{\Lambda\}$, the language whose sole element is the empty string $\{\Lambda\}$:

$$A \cdot \{\Lambda\} = A = \{\Lambda\} \cdot A \quad \text{for all } A \subset X^*.$$

Moreover, $(L_X, \cdot, \{\Lambda\})$ is a noncommutative monoid: in general, $A \cdot B \neq B \cdot A$.

Finally, concatenation distributes over union

$$A \cdot (B \cup C) = A \cdot B \cup A \cdot C$$

for all languages A, B, C over X. Thus, $(L_X, \cup, \cdot, \phi, \{\Lambda\})$ is a semiring. $\square$

EXERCISES FOR SECTION 2.2

1. Use **3**, the inductive definition of X^*, to prove by induction that
    (a) $l(ww') = \ell(w) + \ell(w')$
    (b) $\text{conc}(\text{conc}(w, w'), w'') = \text{conc}(w, \text{conc}(w', w''))$
    (c) $\text{conc}(w, \Lambda) = \text{conc}(\Lambda, w) = w$.

2. Give an *inductive* definition of the "string reversal function"

$$\rho: X^* \to X^* \text{ which sends } x_1 x_2 \cdots x_n \text{ to } x_n \cdots x_2 x_1.$$

3. Prove that the triple $(\mathbf{Z}, +, 0)$ is a monoid.

4. A *semigroup* is a pair (S, m) where S is a set and $m: M \times M \to M$ is associative. (i) Prove that the set of even integers is a semigroup under addition. (ii) Is the set of odd integers a semigroup under addition? Justify your result.

5. Prove that every element of a group has a *unique* inverse.

6. The set $\{0, 1\}$ can be equipped with two operations called *disjunction* and *conjunction*. We shall study these operations in more detail later. Disjunction is denoted by $\vee$, and defined by

$$0 \vee 0 = 0, \qquad 0 \vee 1 = 1 \vee 0 = 1 \vee 1 = 1.$$

Conjunction is denoted by $\wedge$, and defined by

$$0 \wedge 0 = 0 \wedge 1 = 1 \wedge 0 = 0, 1 \wedge 1 = 1.$$

Prove that $(\{0, 1\}, \vee, \wedge, 0, 1)$ is a semiring. It is called the *Boolean semiring*.

7. Let Z_m denote the set $\{0, 1, 2, \ldots, (m - 1)\}$.
   (a) Show that $(Z_m, +_m, 0)$ and $(Z_m, *_m, 1)$ are monoids, where $+_m$ and $*_m$ are addition and multiplication *modulo m*, respectively.
   (b) Show that if m is a prime number, then $(Z_m - \{0\}, *_m, 1)$ is a group.

8. Given two monoids (M_1, m_1, e_1) and (M_2, m_2, e_2) a map $f: M_1 \to M_2$ is called a *homomorphism* from (M_1, m_1, e_1) to (M_2, m_2, e_2) if
   (i) $f(e_1) = e_2$;
   (ii) for all $x, y \in M_1, f(m_1(x, y)) = m_2(f(x), f(y))$.
   If such a map f exists, we also say that (M_2, m_2, e_2) is a *homomorphic image* of (M_1, m_1, e_1), although it may be the case that $f(M_1) \subset M_2$.
   (a) Show that $(Z_m, +_m, 0)$, in Exercise 7, is a homomorphic image of $(\mathbf{N}, +, 0)$, in Example 5.
   (b) Show that the map $f: R \to R, x \mapsto 2^x$ is a homomorphism from the monoid $(R, +, 0)$ to the monoid $(R, \times, 1)$.

9. Describe informally the following languages:
   (a) $\{a, b\}^* \cdot \{a\}^*$
   (b) $\{a, b\}^* \cdot \{aa\}^*$
   (c) $\{a, b\}^* \cdot \{b, a\}^*$
   (d) $\{a, b\}^* \cdot \{a\}^* \cdot \{b\}^*$

2.3 Languages and Automata Theory: A First Look

It is often important in computer science to give two equivalent characterizations of a particular language — one essentially algebraic, the other machine-like. Thus, while a computer language like Pascal has a formal definition, made up of rules for generating all and only legal Pascal programs, at the same time program legality is also characterized by a theoretical machine: a

Pascal compiler. In this section we want to introduce another, similar sort of dichotomy.

We saw in Section 2.2 how union, dot, and star operations may be combined algebraically to form a wide variety of languages over the same finite alphabet. These languages have machine-like characterizations too, and here, as we shall see, the class of machines in question can be described simply and elegantly. We begin with a simple example.

1 Example. The language $\{a^*\} \cdot \{b^*\}$ of Example 2.2.6 is recognized by the following machine:

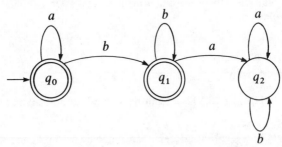

The machine operates as follows on string aab. Starting at position q_0 (indicated by the left-most arrowhead), the machine reads aab from left to right. The first symbol is an "a," and since the arc labeled "a" leading from q_0 returns to position q_0, the machine stays at q_0 as it proceeds to the second symbol. This symbol is also an "a," so the device again returns to q_0 to read the third symbol, a "b." The "b" arrow leads to position q_1. This is the terminal state of the device since the symbol string has been completely scanned. Position q_1 has a double circle around it, indicating that it is an "accepting" state and so the string aab is accepted by this device. Notice that this machine decides "yes" or "no" on every string in $\{a, b\}^*$. The string $aaba$, for example brings the machine to q_2, a rejecting state since it is only circled once. It should be clear that this machine accepts exactly the language $\{a\}^* \cdot \{b\}^*$; the state q_2 can be reached only if an "a" follows a "b" in the input stream of symbols.

The "positions" of our device — q_0, q_1, etc. — are called *states*, and a subset of them, in this case q_0 and q_1, are termed *accepting states*. The state q_0 is called the *start state*. We summarize the structure of these devices with the following definition.

2 Definition. A *finite-state acceptor* (FSA) M with input alphabet X is specified by a quadruple (Q, δ, q_0, F) where

Q is a finite set of states
$\delta: Q \times X \to Q$ is the transition function
$q_0 \in Q$ is the initial state
$F \subset Q$ is the set of accepting states.

3 Example. The FSA of Example 1 is represented by the quadruple

$$(\{q_0, q_1, q_2\}, \delta, q_0, \{q_0, q_1\})$$

where the function δ may be described by means of the following chart:

δ	a	b
q_0	q_0	q_1
q_1	q_2	q_1
q_2	q_2	q_2

4 Example. Let us consider the language $\{a, b\}^* - (\{a\}^*\{b\}^*)$. This language is the complement of the language in Example **1**. Notice that it can be accepted by the same FSA with accepting and nonaccepting states interchanged: $F = \{q_2\}$ instead of $\{q_0, q_1\}$ in this case. Pictorially we have

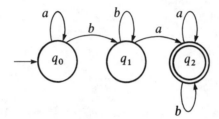

The general idea is that we use the finite-state acceptor M to classify elements of X^* (strings of symbols from the input alphabet X) as follows:

M starts in state q_0, and scans the left-most symbol in some string w in X^*. Based on its current state — in this case q_0 — and the symbol scanned, M shifts to some new state. This transition is governed by the transition function δ given above, which maps a state/symbol pair to a new state. The second symbol is then scanned, with M in the new state, and again a δ-transition occurs. This process continues until the entire input string is read. If the final state is a member of F, then the string x is *accepted* by M. Given an FSA M, we shall denote by $T(M)$ the set of strings accepted by M — the set of $w \in X^*$ which send M from q_0 to some state in F. Our goal is to characterize the set of strings which can be $T(M)$'s for some M.

We now consider $T(M)$ more formally by defining a function $\delta^*: Q \times X^* \rightarrow Q$ such that $\delta^*(q, w)$ is the state which M will go to with an input string w if started in state q_0. We do this by induction:

Basis Step: $\delta^*(q, \Lambda) = q$ for each state q in Q. If M is started in state q, then when it has received the empty input string it must still be in that same state q.

Induction Step: $\delta^*(q, wx) = \delta(\delta^*(q, w), x)$ for each state q in Q, each input string w in X^* and each input symbol x in X.

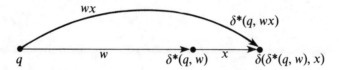

The string w sends M from state q to state $\delta^*(q, w)$, which input x then changes to state $\delta(\delta^*(q, w), x)$ — but this is just the state $\delta^*(q, wx)$ to which the string wx sends M from state q.

5 Definition. The set $T(M)$ of strings accepted by the FSA $M = (Q, \delta, q_0, F)$ with input alphabet X is the subset of X^*

$$T(M) = \{w \,|\, \delta^*(q_0, w) \in F\}$$

comprising all those strings which send M from its initial state q_0 to an accepting state, i.e., a state in F.

We say a subset L of X^* is a *finite state language* (FSL) if it equals $T(M)$ for some FSA M.

EXAMPLES OF FINITE-STATE LANGUAGES

(i) The empty set $\varnothing$ is an FSL since it is $T(M)$ for any M for which the set F of accepting states is empty. Here is the simplest such M (where we take $X = \{0, 1\}$):

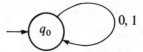

Here $Q = \{q_0\}$, $\delta(q_0, 0) = \delta(q_0, 1) = q_0$, and $F = \varnothing$.

(ii) The following state graph is that of a *parity checker*.

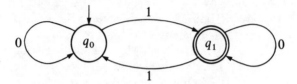

It accepts just those strings with an *odd* number of 1's:

$$\delta^*(q_0, w) = \begin{cases} q_0, & \text{if } w \text{ contains an even number of 1's;} \\ q_1, & \text{if } w \text{ contains an odd number of 1's.} \end{cases}$$

2. What languages do the following FSLs accept?

(i)

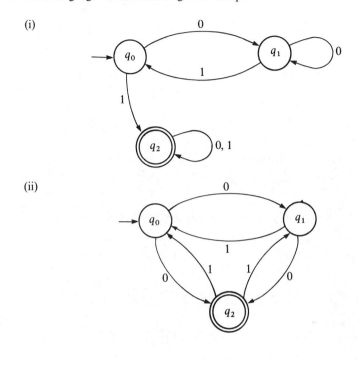

(ii)

2.4 Context-Free Grammars

We learn to parse sentences of English with trees like the one shown in Figure 10. The structure of these rules may be summarized in the form:

$S \rightarrow NP\ VP$ a sentence (S) can take the form of a noun phrase (NP) followed by a verb phrase (VP).

$NP \rightarrow$ Det Adj N a noun phrase (NP) can take the form of a determiner (Det), possibly followed by an adjective (Adj), followed by a noun (N).

$VP \rightarrow V\ NP$ a verb phrase (VP) can take the form of a verb (V) followed by a noun phrase (NP).

Finally, Det $\rightarrow$ the, Adj $\rightarrow$ ugly, $N \rightarrow$ elephant, $N \rightarrow$ petunias, $V \rightarrow$ trampled tell us the parts of speech to which various words of English belong.

Of course, a full grammar of English contains many more replacement rules (e.g., to handle prepositions, negations, conjunctions, relative clauses, etc.) and also contains consistency mechanisms to make sure, for example, that a plural verb takes a plural subject ("they give," not "he give," etc.). Nonetheless, *context-free grammars* — grammars whose grammatical rules are all of the simple replacement kind shown in Figure 10 — have played a

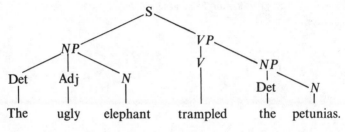

Figure 10

useful role in mathematical linguistics. Grammatical systems of this kind are also extremely important in computer science. This is so because context-free grammars are adequate for expressing much of the *syntax* of programming languages — the rules that determine which strings of symbols constitute well-formed programs.

Consider, for example, describing what strings of symbols constitute a number in decimal notation. Typical such numbers are

179 — a string of digits

.394 — a decimal point followed by a string of digits

15.64 — two strings of digits separated by a decimal point.

We first describe nonempty strings of decimal digits by the grammar

$$M \rightarrow D \,|\, MD$$
$$D \rightarrow 0 \,|\,1\,|\,2\,|\,3\,|\,4\,|\,5\,|\,6\,|\,7\,|\,8\,|\,9$$

The first line says that M may be replaced by D or by M followed by D; the second line says that a D may be replaced by any one of the ten digits. Let us use $w \Rightarrow w'$ to indicate that w' is obtained from w by replacing a single letter with one of its acceptable substituents. Then, for example,

$$M \Rightarrow MD \Rightarrow M9 \Rightarrow MD9 \Rightarrow DD9 \Rightarrow 1D9 \Rightarrow 179$$

Now, we saw that a number could be of the form w, $.w'$, or $w.w'$ where w and w' are nonempty strings of digits. We can describe this by

$$N \rightarrow M \,|\, .M \,|\, M.M$$

A derivation of 15.64 from N is then given by

$$N \Rightarrow M.M \Rightarrow MD.M \Rightarrow MD.MD \Rightarrow DD.MD \Rightarrow DD.DD$$

$$\Rightarrow 1D.DD \Rightarrow 15.DD \Rightarrow 15.6D \Rightarrow 15.64$$

which corresponds to the derivation tree shown in Figure 11.

The above example is a special case of the following:

1 Definition. A context-free grammar $G = (V, T, S, P)$ is specified by a finite set V of symbols called the *nonterminals* (or *variables*), a finite set T

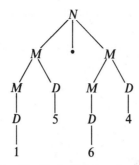

Figure 11   Derivation tree for 15.64.

of symbols, disjoint from V, called *terminals*, an element S of V called the start symbol, and a finite subset P of $V \times (V \cup T)^*$ called the set of *productions*.

In the above example, we have a grammar G_1 with

$$V = \{N, M, D\}$$
$$T = \{., 0, 1, 2, 3, 4, 5, 6, 7, 8, 9\}$$
$$S = N$$

$$P = \{(N, M), (N, .M), (N, M.M), (M, D), (M, MD), (D, 0), \ldots, (D, 9)\}$$

(Note then that we write $v \to w_1 | w_2 | \cdots | w_n$ if $(v, w_1), (v, w_2), \ldots, (v, w_n)$ are all the productions in P with v in V as first element.) Then the decimal numbers are precisely those strings which can be derived from N (the start symbol of this grammar) and which only contain terminal symbols, i.e., the decimal point and the digits. This generalizes as follows.

2 Definition. Let $G = (V, T, S, P)$ be a context-free grammar. We define the binary relation $\Rightarrow$ on $(V \cup T)^*$ — $w \Rightarrow w'$ is read as w *directly derives* w' — by specifying that

$w \Rightarrow w'$ just in case w' is obtained from w by replacing a single variable in a fashion specified by a production in P, i.e., just in case there exist w_1, w_2 in $(V \cup T)^*$, v in V and $(v, \bar{w})$ in P such that $w = w_1 v w_2$, $w' = w_1 \bar{w} w_2$.

Note that $\Rightarrow$ is then a binary relation on $(V \cup T)^*$. Section **1.3** showed how to associate with any relation $R: A \to A$ its reflexive transitive closure $R^*: A \to A$. In the same way, we define the binary relation $\overset{*}{\Rightarrow}$ on $(V \cup T)^*$ as the reflexive transitive closure of $\Rightarrow$. $w \overset{*}{\Rightarrow} w'$, read as w *derives* w', just in case $w = w'$ or there exists a sequence $w_1, w_2, \ldots, w_n$ ($n \geq 2$) of strings in $(V \cup T)^*$ such that

$$w = w_1;$$
$$w_j \Rightarrow w_{j+1} \quad \text{for } 1 \leq j < n;$$
$$w_n = w'.$$

3. The language $L(G)$ generated by G is the set of terminal strings which can be derived from S:

$$L(G) = \{w \,|\, w \in T^*, S \overset{*}{\Rightarrow} w\}.$$

We say that a set $A \subset T^*$ is a *context-free language* just in case it is an $L(G)$ for some context-free grammar G.

Here is a second example of a context-free grammar, $G_2 = (V, T, S, P)$:

$$V = \{S\}$$

$$T = \{a, b\}$$

$$S \rightarrow aSb \,|\, ab.$$

We see that the only possible derivations are

$$S \Rightarrow ab$$

$$S \Rightarrow aSb \Rightarrow aabb$$

$$S \Rightarrow aSb \Rightarrow aaSbb \Rightarrow aaabbb,$$

etc. Thus, if we use the notation a^n to abbreviate a string of n occurrences of the letter a,

$$L(G) = \{a^n b^n \,|\, n \geq 1\}.$$

Let us consider a third example. A *palindrome* is a string of characters, like "pop," "madam," or "otto," which is the same forwards and backwards. Odd length palindromes such as "madam" pivot about a unique middle character; even length palindromes — "otto" — simply divide in half. There are a number of famous palindromes in English (blanks ignored), such as "A man a plan a canal Panama" and "Able was I ere I saw Elba." One of the longest English palindromes (ignoring spaces and periods!) is "Doc note I dissent. A fast never prevents a fatness. I diet on cod." If we restrict our attention to a two letter terminal alphabet, say $T = \{a, b\}$, then we can easily write a context-free grammar generating all and only palindromes over this character set:

4. $S \rightarrow aSa \,|\, bSb \,|\, a \,|\, b \,|\, \Lambda.$

[Here, and in what follows, we just specify the productions when the rest of the grammar can be understood.]

The derivation of *ababa* proceeds as follows:

$$S \Rightarrow aSa \Rightarrow abSba \Rightarrow abaSaba \Rightarrow ababa.$$

How do we know for sure that this grammar (i) derives only palindromes; and (ii) derives every palindrome over $\{a, b\}$? The assertion that the grammar above meets its *specification* is a claim that requires formal proof. We now consider how such proofs may be obtained, and in particular we look at the close connection between context-free grammars and proofs by induction.

INDUCTION REVISITED

Let us reconsider the grammar with productions

$$S \to aSa \,|\, bSb \,|\, a \,|\, b \,|\, \Lambda.$$

A proof that this grammar generates all and only palindromes is easily accomplished using induction. In fact, the very form of the productions in the grammar supplies us with exactly the facts we need to make the proof work.

In order to prove this result, however, it will be useful to appeal to a slightly extended principle of induction, which can be shown to be equivalent to the induction principle developed in Section 2.1.

5 The Principle of Induction Extended. In order to prove a property $P(n)$ for all $n \in \mathbf{N}$, it is sufficient to prove

(i)   a *Basis Step*: $P(0)$ is true; and
(ii) an *Induction Step*: If $P(k)$ holds for all $k \le n$, then $P(n + 1)$ is true.

It goes without saying that this principle can be used to prove results by induction with basis case $m > 0$ as well.

Notice how this principle, which is sometimes called *complete induction*, differs from our earlier notion of inductive proof. Here the induction step allows the assumption of $P(k)$ for all $k < n$ in addition to the assumption $P(n)$.

Using this extended principle of induction, we can now prove that our palindrome grammar does indeed meet its specification.

6 Lemma. *$L(G)$ for the grammar G whose productions are given in **4** is just the set of palindromes over $\{a, b\}$.*

PROOF (i) We first show that our grammar generates only palindromes.

Basis Step: Suppose a string $x \in L$, and $\ell(x) = 0$ or $\ell(x) = 1$. Strings of length 1 or 0 may only be derived by the productions

$$S \to a, \quad S \to b, \quad S \to \Lambda.$$

Hence, for strings of these lengths, our (basis) claim that only palindromes are produced is indeed correct.

Induction Step: Next, suppose that for all strings $x \in L(G)$ such that $\ell(x) \le n$, where $n \ge 1$, x is a palindrome. We must show that if $y \in L(G)$, $\ell(y) = n + 1$, then y is a palindrome. We argue as follows: If $S \overset{*}{\Rightarrow} y$, $\ell(y) \ge n$, then the derivation of y must begin with either

$$S \to aSa \quad \text{or} \quad S \to bSb.$$

For definiteness, let us assume the first case holds (the argument for the second case is exactly the same). Thus

$$S \Rightarrow aSa \overset{*}{\Rightarrow} y$$

and hence y begins and ends with an a. But then the S occurring between the a's in the center expression must derive a string y' of length $n - 1$, where $y = ay'a$. By our inductive hypothesis, y' is a palindrome, and therefore so is y, since pasting an "a" at the front and back of a palindrome preserves the symmetry of string reversal. Thus our grammar generates only palindromes.

(ii) A similar argument establishes that our grammar generates every palindrome. As our basis step, we consider all length 0 and length 1 palindromes over $\{a, b\}$. There are only three palindromes of these lengths, Λ, a, b, and G trivially generates all of these. Now suppose G generates all palindromes of length $\leq n$, for $n \geq 1$. Let x be a length $n + 1$ palindrome. We must use the inductive hypothesis and the actual productions of the grammar to show how x may be generated. The first and last characters of x must be the same, and without loss of generality, let us assume this character is "a." That is, $x = ax'a$, where x' is a string of length $n - 1$. Since x' falls in the "center" of x, x' must itself be a palindrome. Hence, by the inductive hypothesis $(\ell(x') \leq n) S \overset{*}{\Rightarrow} x'$. But then x is derivable by the sequence $S \Rightarrow aSa \overset{*}{\Rightarrow} ax'a = x$.   □

Consider, again, the context-free grammar which generates all the decimal numbers: $N \rightarrow M | .M | M.M$ with $M \rightarrow D | MD$. If D stands for any of the ten digits, the $L(G_1)$ of this grammar is the set of strings accepted by the machine [Exercise: Restructure the machine to avoid the two D-transitions from q_0.]:

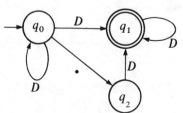

q_0 is the only initial state; q_1 is the only accepting state; and the set of accepted strings is just $L(G_1)$.

This raises the question: Is every context-free language a regular set? The answer is no, as the following observation shows.

7 Observation. *The language* $\{a^n b^n | n \geq 1\}$ *is not accepted by any finite state acceptor M.*

PROOF. Suppose that every string of the form $a^n b^n$ were accepted by M. We shall show that M must accept a string not of the form $a^n b^n$. Let M have k states, and consider the sequence

$$(\delta^*(q_0, a^n): 0 \leq n \leq k)$$

where we use $\delta^*(q_0, a^n)$ to denote the state of M reachable from the initial state q_0 by applying a string n a's. There are $k + 1$ states in this sequence, and so at least two of them must be the same, say

$$\delta^*(q_0, a^r) = \delta^*(q_0, a^s) \quad \text{for some } 0 \le r < s \le k.$$

But then as we see from the diagram

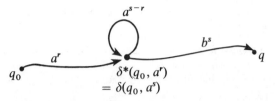

we must have that

$$\delta^*(q_0, a^r b^s) = \delta^*(\delta^*(q_0, a^r), b^s)$$

$$= \delta^*(\delta^*(q_0, a^s), b^s) = \delta^*(q_0, a^s b^s)$$

which belongs to F, since M accepts every string in $\{a^n b^n : n \ge 1\}$. Thus M accepts $a^r b^s$. But $r \neq s$, and so M cannot accept the exact set $\{a^n b^n | n \ge 1\}$. □

The detailed study of context-free languages must be left to another volume. We close our present discussion with an example of an alternative notation for context-free grammars which is popular for specifying the syntax of programming languages. We present the syntax of decimal numbers in the form:

⟨decimal number⟩ ::= ⟨unsigned integer⟩ | · ⟨unsigned integer⟩ |

⟨unsigned integer⟩ · ⟨unsigned integer⟩

⟨unsigned integer⟩ ::= ⟨unsigned integer⟩⟨digit⟩ | ⟨digit⟩

⟨digit⟩ ::= 0|1|2|3|4|5|6|7|8|9

Each nonterminal is represented by a name in angle brackets ⟨ ⟩; we use ::= instead of the replacement arrow; and we use | to separate the alternative replacements for the nonterminal occurring to the left of the ::= in the display. This way of representing the productions of a context-free grammar was introduced by Backus and Naur in specifying the syntax of the programming language Algol, and is referred to as BNF for Backus–Naur Form.

8 Example. The inductive definition of X^* can be recast in the BNF as follows:

⟨string⟩ ::= Λ | ⟨element⟩⟨string⟩

Here elements of X^* correspond to "string" and elements of X correspond to "element."

EXERCISES FOR SECTION 2.4

1. Write context-free grammars for the following languages:
   (a) $L_1 = \{a^n b^m a^n \mid n, m \geq 1\}$ (A typical string: $aaabbaaa$)
   (b) $L_2 = \{a^n b^m cb^m a^n \mid n, m \geq 1\}$ (A typical string: $a^2 b^3 cb^3 a^2$)
   (c) $L_3 = \{x \in \{a, b\}^* \mid x \text{ contains an equal number of } a\text{'s and } b\text{'s}\}$
       (A typical string: $baaababb$).

2. Suppose L_1 and L_2 are context-free languages over the alphabet $V = \{a, b\}$. Prove that $L_1 \cup L_2$ is also a context-free language.

3. Suppose L is a context-free language over $\{a, b\}$. Define L^R to be the set of strings obtained by reversing all the strings in L. Show that L^R is also context-free. (Hint: What change in the productions of a grammar for L might bring this about?)

4. Assign parts of speech to the words of the sentence, "John walked the salty dog," and then provide a parse tree for it using the grammar presented at the beginning of this section.

5. (a) What language does the grammar

$$S \to aaA$$

$$A \to aa \mid aaA \mid B$$

$$B \to b \mid bB$$

   generate? Prove your result.
   (b) Is this language an FSL?

6. Use the technique of Observation 7 to show that the following languages cannot be accepted by any finite state acceptor.
   (a) $\{a^n b^m \mid n \geq m\}$
   (b) $\{a^n b^m \mid n \leq m\}$
   (c) $\{a^k \mid k \text{ is a power of 2}\}$.

2.5 Processing Lists

We may regard the typical element of X^* as being a *list*, $(x_1, x_2, \ldots, x_n)$, whose typical entry x_j is an element of X. In this section we look at a typical data-retrieval problem — finding someone's number in a telephone directory — as an exercise in list-processing. We look at it in three passes: first, informally and set-theoretically (and with some comments on Pascal, an increasingly popular programming language); second, using the inductive definition of X^*; and, third, in the context of an introduction to the List Processing language, Lisp. The algorithms in this section will involve *sequential search* — looking at one entry after another till we find the one we want. In Section 3.3 we shall look at a different method, called *binary search*, which yields a far more efficient algorithm.

A telephone directory can be viewed as a finite list

$$(\text{entry}_1, \text{entry}_2, \ldots, \text{entry}_k, \ldots, \text{entry}_n)$$

of entries. Each entry is in turn a list of three elements

$$\text{entry}_k = (\text{name}_k, \text{address}_k, \text{number}_k).$$

These three items are themselves lists of characters. For the present purpose, we need not analyze the character set, but it is worth noting that it must include a blank character, which we denote '' (quotes with a single space between them). The blank is *not* the empty string (hitting the space bar on a typewriter is not the same as hitting no key at all): $\ell('') = 1$, $\ell(\Lambda) = 0$, and

BOGGS''E''Q = BOGGS E Q $\neq$ BOGGSEQ = BOGGSΛEΛQ

In any case, we define the format for a telephone directory using a character set called *character* (we use mnemonic names in italics to denote sets here, rather than the single letters that prove convenient in more general algebraic settings). We set up three subsets of *character**:

  name = set of character strings which serve as names,

  address = set of character strings which serve as addresses,

  number = set of character strings which serve as phone numbers.

For example, a typical element of *name* might consist of a family name followed by a blank followed by a string of initials separated by blanks; while a typical element of *number* might have the form 555-1212, three digits followed by a hyphen and four more digits, so that *number* = $D^3 \times \{-\} \times D^4$.

We then define *entry*, the set of possible entries, as a Cartesian product

$$entry = name \times address \times number.$$

The set of possible directories is then the set of sequences of entries,

$$directory = entry^*.$$

A directory W is an element of *directory*, i.e., a finite (possibly empty) list of entries, each of which is an element of *entry*. (In the present discussion we have *not* required that the directory be arranged in alphabetical order by name, or that an entry occur only once. We shall look at a program which searches the directory item-by-item until the desired entry is found. In a course on data structures you will learn about *sorted* lists. Search time for sorted lists can be cut by making intelligent use of the knowledge of whether the desired item occurs before or after the one under current scrutiny. See Section 3.3 for an example of this.)

Here, then, is an informal program to determine the phone number of BOGGS E Q in a directory by a *sequential search* (going through the list one entry at a time). The program will give NIL as output if there is no entry under this name.

1. Let the first entry be the current entry.
2. Is BOGGS E Q the name in the current entry?
   if YES: Print out the number of that entry, and HALT.
   if NO: Is there another entry left in the directory?
               if YES: Make it the current entry and return to 2.
               if NO: Print out NIL and HALT.

ARRAYS AND PASCAL

We now show how this can be written in a Pascal-like language. We start by specifying the number N of entries in a directory, so that we require that each directory belong to the N-fold *Cartesian power*

$$directory = entry^N \qquad (1)$$

rather than having the arbitrary length allowed members of *entry**.
Each entry is still a list of three elements, so we have the *Cartesian product*

$$entry = name \times address \times number. \qquad (2)$$

Before giving the Pascal program, we must show how Pascal represents the definitions (1) and (2). In a programming language, a *data type* is a set together with certain basic operations that a program may perform upon it. Let us assume that we have already been given the data types *name*, *address*, and *number* and that we have operations available which let us test two strings of the same data type to see if they are equal. In carrying out the definitions (1) and (2), we not only want to build up the right set of elements, but wish to have operations for reading out specified elements of the list. For example, in (1) we need N maps, *directory* → *entry*, $w \mapsto w[j]$ which let us read out the *j*th entry of the directory w, $1 \le j \le N$. And in (2) we need three *selector* maps

$$entry \rightarrow name, \qquad x \mapsto x.sname$$

$$entry \rightarrow address, \qquad x \mapsto x.saddress$$

$$entry \rightarrow number, \qquad x \mapsto x.snumber$$

where *x.sname*, *x.saddress*, and *x.snumber* are, respectively, the first, second, and third component of *x*. If, for example, $x = $ (BOGGS E Q, 99 WILLOW RD, 555-1212), then $x.snumber = $ 555-1212.
Pascal uses the notation (known as a data type declaration)

type *entry* = **record** *name*: a;
                              address: b;
                              number: c
                    end;

to indicate that the new data type *entry* is the *Cartesian product* of three already-defined data types *a*, *b*, and *c*, and that the respective entries in an item of this Cartesian product can be accessed by using the selectors *sname*, *saddress*, and *snumber* respectively.

Pascal uses the notation (data type declaration)

type *directory* = **array** $[1..N]$ **of** *entry*;

to indicate that the new data type *directory* is the *N*-fold Cartesian power of the already defined data type *entry*, and that the *j*th entry $(1 \le j \le N)$ in an item of this Cartesian power can be accessed by using the maps $w \mapsto w[j]$.

A Pascal program to scan our directory would then look something like this (where we use the braces { } to enclose comments that are not part of the program):

        type *entry* = **record** *name*: *a*;
                     address: *b*;
                     number: *c*
           end;
        type *directory* = **array** $[1 .. N]$ **of** *entry*;
        var: *w*: directory;
          x: entry;
          nm: *a*;
          nr: *c*;
          i: $1..N$;

{So far, we have just repeated the definition of *entry* in terms of *a*, *b*, and *c*, and of *directory* in terms of *entry*. We have then introduced five variables, and have specified what data types each will take its values from for the program. For example, *nr* will take its value from *c*, i.e., each value will be a phone number. The last line tells us that *i* can take any value in the range 1 through *N*.}

load(*w*), *load*(*nm*)

{This is not formal Pascal notation, but indicates that the program must assign to *w* as value the actual directory that is to be searched; and must assign to *nm* the name for which we wish to find the corresponding phone number.}

begin for *i* := 1 **to** *N* **do**

{This line tells the computer to repeat the operations below, starting first with *i* = 1, and then repeating the procedure, increasing *i* by 1 each time until either the **goto** 1 sends the computation to the line of the program labelled 1; or until the program has completed the entire procedure *N* times. In the latter case, control is then transferred to the next line of the program.}

    begin *x* := *w*[*i*]; {get the *i*th entry of the directory *w*}
        if *x.sname* = *nm* **then begin** *nr* := *x.snumber*; **goto** 1 **end**
    end;
    nr := "NIL";

{Set *nr* to the *number* item *x.snumber* of the first entry $x = w[i]$ whose *name* item *x.sname* equals the given name *nm*. If no such entry is found, set *nr* to "NIL." (The quotes indicate that it is the actual string NIL of three characters that is required.)}

       1 Print ("THE PHONE NUMBER FOR" *nm* "IS" *nr*)

end

In a flow diagram, leaving aside the declarations:

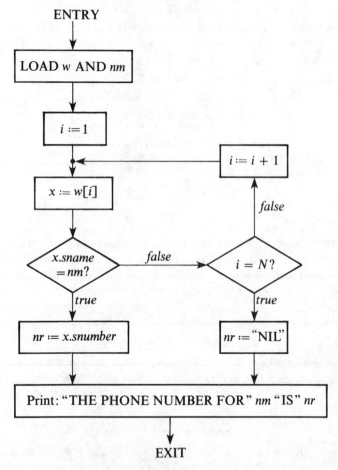

We have just seen how to represent a directory as an *array* of fixed length, and then use a *counter* to keep track as we work our way through the directory, one entry at a time. This, then, is the *iterative* solution to the phone number problem. We now return to the *inductive definition* of X^* and see how this allows us to define the phone-number-retrieval function by induction.

    Our goal, then, is to define the function PHONENR: *directory* × *name* → *number* so that PHONENR(*w, nm*) is the phone number occurring in the first entry of w whose name item is nm; and is NIL if no such item occurs in w. This immediately yields the

Basis Step: PHONENR(Λ, nm) = NIL
for if there are no entries in the directory ($w = \Lambda$), there are certainly no entries which start with nm. For our induction step, note that a nonempty directory w has the form (x, w') with x in *entry* and w' in *directory*. Then either x starts with nm, and we are done, or we simply forget about x, and move on to search w'. We thus have the

Induction Step: If $w = (x, w')$ with x in *entry* and w' in *directory*, then

(a)   If $x.sname = nm$,   then PHONENR($(x, w'), nm$) = $x.snumber$
(b)   If $x.sname \neq nm$,   then PHONENR($(x, w'), nm$)

$$= \text{PHONENR}(w', nm).$$

s-EXPRESSIONS AND LISP

We now introduce a few basic features of the programming language Lisp, and show that they permit us to translate the inductive definition of a function such as PHONENR directly into a Lisp program, without the use of counters.

We first introduce the basic Lisp data structure, the *s-expression*, (short for *symbolic-expression*). In defining strings in X^*, our inductive step was to form (x, w) where x was in X and w was a previously formed element of X^*. By contrast, the inductive step in the formation of s-expressions in Lisp combines a *pair* of s-expressions.

1 Definition. Let A be a set. Then the set $S(A)$ of *s-expressions over* A is defined inductively by:

Basis Step: Each a in A belongs to $S(A)$.
Induction Step: If w_1 and w_2 are s-expressions, then the *dotted pair* $(w_1 \cdot w_2)$ is also an s-expression.
We say that an element of $S(A)$ is an *atom* just in case it belongs to A.

Lisp further requires that A contain two distinguished elements, T and NIL. These two elements are used to denote the truth values *true* and *false*, respectively, so that a predicate is represented in Lisp by a function $S(A) \to \{T, \text{NIL}\}$. NIL also plays a second role, namely the role that the symbol Λ played in the inductive definition of X^*. (Warning: Different computer centers have different versions of Lisp. In what follows, we present basic concepts of Lisp, but we do not expect that the programs will run as printed. However, with the understanding that this section provides, you should find it fairly easy to use a manual to get programs running if your neighborhood system has a Lisp interpreter.)

2 The basic predicate of Lisp is ISATOM: $S(A) \rightarrow \{T, \text{NIL}\}$, which tells whether or not an s-expression is an atom

$$\text{ISATOM}(w) = \begin{cases} T, & \text{if } w \text{ is in } A; \\ \text{NIL}, & \text{if not.} \end{cases}$$

An s-expression can also be represented by a Lisp or L-tree. For example, the s-expression $((a \cdot b) \cdot c)$ is represented by the L-tree

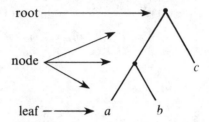

We see that an L-tree is a collection of straight lines. The lines are called *branches* while the ends of the lines are called *nodes* or *vertices*. We see that there is one node at the top called the *root*, and that every other node is at the lower end of a single branch. Each node is either at the top of two branches (one to the left and one to the right), or it is at the top of no branches at all. In the latter case we call the node a *leaf*, and label it with an atom. We say that a leaf is a *terminal* node, and that other nodes are *nonterminal*.

The rule for representing an s-expression by a tree is easily given inductively:

Basis Step: If a is an atom, then the tree $\mathbf{T}(a)$ representing a comprises a single node labelled by a.

Induction Step: If w_1 and w_2 are s-expressions represented by the trees $\mathbf{T}(w_1)$ and $\mathbf{T}(w_2)$ respectively, then $(w_1 \cdot w_2)$ is represented by

$\mathbf{T}(w_1) \qquad \mathbf{T}(w_2)$

For example, $\mathbf{T}((a \cdot b) \cdot (a \cdot (b \cdot c)))$ equals

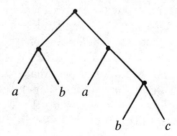

To see how s-expressions may be represented in the computer, we may redraw the tree by replacing each non-terminal node by a box

and each terminal labeled a by a box

We can then rearrange the position of the nodes without losing track of what comes first. In this *box-notation*, $((a \cdot b) \cdot (a \cdot (b \cdot c)))$ becomes

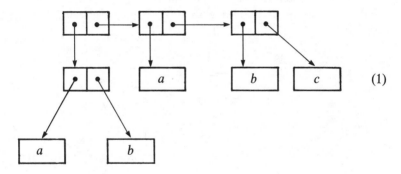

(1)

One strategy for storing s-expressions in a computer has each box correspond to a word in computer memory. Words in one portion of memory — say, words in a particular range of memory addresses — are treated as atoms. Words outside this range correspond to nonatomic s-expressions, and are divided in two. Each half of a "nonatomic" word contains the *address* of the word toward which it points. Notice that with this scheme, ISATOM (recall **2**) is particularly easy to compute — the address of ISATOM's argument determines if the argument is an atom.

Given a dotted pair $(w_1 \cdot w_2)$, it is important to be able to extract w_1 and w_2. Conversely, given w_1 and w_2 we want to be able to form them into $(w_1 \cdot w_2)$. We thus need three functions:

1. HEAD: $S(A) \rightarrow S(A)$ is a *partial* function defined inductively by

$$\begin{cases} \text{HEAD}(a) = \bot \text{ (i.e., is undefined)} & \text{if argument } a \text{ is an atom,} \\ \text{HEAD}((w_1 \cdot w_2)) = w_1 & \text{otherwise.} \end{cases}$$

2. TAIL: $S(A) \rightarrow S(A)$ is a *partial* function defined inductively by

$$\begin{cases} \text{TAIL}(a) = \bot & \text{if argument } a \text{ is an atom,} \\ \text{TAIL}((w_1 \cdot w_2)) = w_2 & \text{otherwise.} \end{cases}$$

3 CONS: $S(A) \times S(A) \to S(A)$ is the *construction* function defined explicitly by

$$\text{CONS}(w_1, w_2) = (w_1 \cdot w_2).$$

In Lisp, HEAD is called CAR, and TAIL is called CDR. It may help to notice that A comes before D, so CAR comes before CDR. But to understand the notation, we must return to the box notation (1) for $w = ((a \cdot b) \cdot (a \cdot (b \cdot c)))$. If the interpreter has the address α for the root node of w, it obtains the address of the root node of $\text{CAR}(w) = (a \cdot b)$ simply by reading the first half of the contents of the word at address α. Now, in the IBM 7090, the machine on which Lisp programs were first written, the first half of a word of computer memory was called the Address Register — and so CAR was short for *Contents* of the *Address Register*. The second half of a word in memory was called the *Decrement Register* — and thus CDR.

We can of course compose the above functions. For example, it is always true that

$$\text{CAR}(\text{CONS}(w_1, w_2)) = w_1$$

and

$$\text{CDR}(\text{CONS}(w_1, w_2)) = w_2.$$

Turning to a specific example

$$\text{CAR}(\text{CDR}(\text{CDR}((a \cdot (b \cdot (d \cdot c)))))) = \text{CAR}(\text{CDR}((b \cdot (d \cdot c))))$$
$$= \text{CAR}((d \cdot c))$$
$$= d.$$

Since this use of chains of CARs and CDRs is very common in Lisp, there is a standard abbreviation — write C, then the middle letter of each CAR and CDR in order, then close with R. So, for example, CADDR is short for CAR · CDR · CDR, and we have

$$\text{CADDR}((a \cdot (b \cdot (d \cdot c)))) = d$$

while

$$\text{CADR}((a \cdot (b \cdot (d \cdot c)))) = b$$

and

$$\text{CDAR}((a \cdot (b \cdot (d \cdot c)))) = \bot.$$

Lisp has a special function EQ which can compare two atoms and tell if they are equal:

$$\text{EQ}(w_1, w_2) = \begin{cases} T & \text{if } w_1 \text{ and } w_2 \text{ are both atoms and } w_1 = w_2, \\ \text{NIL} & \text{otherwise.} \end{cases}$$

Let us use this to define a Lisp program to check whether two arbitrary s-expressions are equal. (Repeated warning: Your local Lisp interpreter may require a different format from that used here. But the ideas are the same.)

We have to define EQUAL: $S(A) \times S(A) \to \{T, \text{NIL}\}$ so that

$$EQUAL(w, w') = \begin{cases} T & \text{if } w = w' \\ \text{NIL} & \text{if } w \neq w'. \end{cases}$$

This is easy if w is an atom, since then

$$EQUAL(w, w') = EQ(w, w').$$

If w is *not* an atom but w' *is* an atom, we have

$$EQUAL(w, w') = \text{NIL}.$$

But if neither w nor w' is an atom, then we have

$$w = (w_1 \cdot w_2) \quad \text{where } w_1 = CAR(w) \text{ and } w_2 = CDR(w)$$
$$w' = (w'_1 \cdot w'_2) \quad \text{where } w'_1 = CAR(w') \text{ and } w'_2 = CDR(w')$$

and it is clear that

if and only if both $w_1 = w'_1$ and $w_2 = w'_2$.

The Lisp conditional generalizes the **if**...**then**...**else** construct that we have already used. Its general form is

$$(COND(P_1 \quad F_1)$$
$$(P_2 \quad F_2)$$
$$\vdots$$
$$(P_n \quad F_n))$$

and it is executed as follows: If the predicate P_1 is true, then evaluate F_1 to obtain the result; otherwise, if P_2 is true, evaluate F_2 to obtain the result; ...; if P_n is true, evaluate F_n to obtain the result; if none of P_1 through P_n is true, then the result is undefined.

If P_n is just T, then should none of P_1 through P_{n-1} evaluate to T, then F_n will automatically be evaluated to obtain the result. We use AND to combine two truth values.

$$AND(T, T) = T, \text{ while}$$

$$AND(T, \text{NIL}) = AND(\text{NIL}, T) = AND(\text{NIL}, \text{NIL}) = \text{NIL}.$$

Thus we have the following recursive Lisp program for EQUAL:

EQUAL(w, w') = (COND((ISATOM(w)) (EQ(w, w')))
                              ((ISATOM(w')) NIL)
                              (T ((EQUAL(CAR(w), CAR(w')))AND
                              (EQUAL(CDR(w), CDR(w'))))))).

Look at how EQUAL compares the L-trees

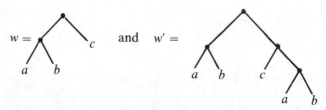

Since the first two tests (ISATOM(w) and ISATOM(w')) both return NIL,
we have to test EQUAL(CAR(w), CAR(w')) and EQUAL(CDR(w),
CDR(w')). The first chore reduces to testing both

EQUAL(a, a)   and   EQUAL(b, b)

which both return T, so that EQUAL(CAR(w), CAR(w')) = T. However,
CDR(w) = c is an atom, and so

EQUAL(CDR(w), CDR(w')) = EQ(c, CDR(w'))

= NIL, since CDR(w') is not an atom.

Thus

EQUAL(w, w') = EQUAL(CAR(w), CAR(w'))AND EQUAL(CDR(w),
               CDR(w'))

= T AND NIL

= NIL

and so our program confirms (correctly)that $w \neq w'$ in this case.

So far in our discussion of Lisp we have considered as objects of study
only s-expressions — trees built up from atoms using the operator CONS.
But clearly a much more reasonable representation for many objects in
computer science is a list: ($w_1 w_2 \cdots w_n$). Now lists are such a common and
important data structure that Lisp systems permit list notation, with the
understanding that a list — say ($A\, B\, C$) — is an abbreviation for the s-
expression ($A \cdot (B \cdot (C \cdot \text{NIL}))$), i.e.,

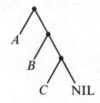

Notice that $(CAR(A\ B\ C)) = A$, $CDR(A\ B\ C) = (B\ C)$, and $CONS(A\ (B\ C))'$ $= (A\ B\ C)$.

Now let us return to our telephone directory. Using list notation, we can represent our directory as

$$(w_1 w_2 \cdots w_{n-1} w_n) \tag{2}$$

and a typical entry of the telephone book is of the form

$$e = (\alpha\ \beta\ \gamma) \tag{3}$$

where α is a name, β is an address, and γ is a number.

Now at the end of the previous subsection we defined

$$PHONENR: directory \times name \rightarrow number$$

recursively by

Basis Step: $PHONENR(\Lambda, nm) = NIL$
Induction Step: Let $w = (x, w')$. If $nm = x.sname$ then

$PHONENR(x, nm) = x.snumber$, else $PHONENR(w, nm)$
$= PHONENR(w', nm)$.

We can immediately transform this into a Lisp program which will take any directory w that is legally defined as a list of the form (2) whose entries are legally defined as (name, address, number) triples of the form (3), and any name nm and will return the third component of the first entry for which nm is the first component; and will return NIL if no such entry exists in w. Here is our phone number program in Lisp. Notice that if $CAR(w)$ is the first entry in the directory, then $CAAR(w)$ is the name field of the first entry, and $CADDR(w)$ is the number field of the first entry.

$PHONENR(w, nm) = (COND(EQUAL(NIL, w)\ NIL)$
$(EQUAL(nm, CAAR(w))\ CADDAR(w))$
$(T\ PHONENR(CDR(w), nm)).$

EXERCISES FOR SECTION 2.5

1. Suppose a university keeps a record of each student which contains in order, the student's name, address, year of graduation, and — for up to 50 courses — the course number, title, semester of study, and final grade. Provide an explicit description of a Cartesian product of sets to which each record belongs. Provide a data type declaration for this record structure. (Hint: To make all records the same length, include in the set of course numbers a "dummy number," call it NOTYET, to fill in for courses not yet taken.)

2. Write down an iterative program to compute the grade-point average for a student from the record of Exercise 1. (Let each course grade be a number between 0 and 4. Sum up the course grades, then divide by the number of courses already taken.)

3. Define a recursive structure which is a list of course results, each of which comprises a course number and final grade for a course. Provide a recursive program which can compute the grade-point average (see Exercise 2) from such a list.

4. Write the following s-expressions and lists in box notation.
   (i) $(a \cdot (b \cdot (c \cdot c)))$
   (ii) $(A\ B\ C\ D)$
   (iii) $(A\ B\ C(D\ E))$

5. What is CADAR$((A\ B)\ C\ D\ E)$?

6. Describe in words what the following Lisp program does.

$$\text{FIRST}(S) = (\text{COND}(\text{ATOM}(S)\ S)$$

$$(T\ \text{FIRST}(\text{CAR}(S)))).$$

7. Let A be an atom, and B be an s-expression. Provide a recursive definition of a function IS.MEMBER such that IS.MEMBER(A, B) equals T if A occurs in the s-expression B, but equals NIL otherwise.

CHAPTER 3

Counting, Recurrences, and Trees

Section 3.1 introduces a number of principles useful in counting complicated sets of objects. These principles include the rule of sum, the rule of product, and the pigeonhole principle. The section also introduces permutations and combinations, and the binomial theorem. Section 3.2 extends the formal study of trees, which were briefly encountered in Chapter 2, and explores ways of counting and computing based on recurrence relations. The final section has a different flavor but is still devoted to counting — this time to counting the number of steps taken by an algorithm to process data. It thus introduces the reader to the important topic of "analysis of algorithms" which enables us to compare the efficiency of different approaches to solve a given problem.

3.1 Some Counting Principles

Counting related objects in large collections is a matter of finding a systematic way for listing the members of the collection. Some of these ways occur frequently enough in practice so that we may formulate them as "counting rules" or "counting principles."

1 The Rule of Sum. *Let X and Y be mutually exclusive events (that is, X and Y cannot both occur at the same time). If X can happen in m different ways, and Y can happen in n different ways, then the event (X or Y) can happen in (m + n) different ways.*

2 Example. Suppose that statement labels in a programming language are either single alphabetic symbols or single decimal digits. If we call X the event that a statement label is an alphabetic symbol, then X can happen in any of 26 different ways, i.e., X can take any value from the set $\{A, B, C, \ldots, Z\}$. If we call Y the event that a statement label is a decimal digit, then Y can happen in any of 10 different ways, corresponding to the distinct choices of values, from the set $\{0, 1, 2, \ldots, 9\}$. Hence a statement label, whether alphabetic symbol or decimal digit, denoted by the event $(X$ or $Y)$ can happen in $(26 + 10) = 36$ possible different ways.

3 The Rule of Product. *Let X and Y be mutually exclusive events. If X can happen in m different ways and Y can happen in n different ways, then the event $(X$ and $Y)$ can happen in $m \times n$ different ways.*

4 Example. If we can go from Amherst to Boston in 3 different ways, say $\{a, b, c\}$ and from Boston to New York in 5 different ways, say $\{1, 2, 3, 4, 5\}$, then we can go from Amherst to Boston *and* then from Boston to New York in $3 * 5 = 15$ different ways. This situation can be described by the tree of Figure 12.

Thus the different ways of going from Amherst to New York *via* Boston may be specified by the set $\{a1, a2, a3, \ldots, c4, c5\}$ — altogether 15 different routes.

The reader may recognize these rules as being new versions of facts about the number of elements in a set given in Fact **1.1.2**. If A is the set of different ways in which X can occur, and B is the set of different ways in which Y can occur, then the rule of sum asserts that

$$|A \cup B| = |A| + |B|$$

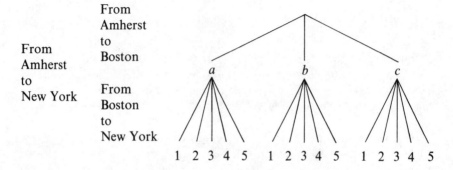

Figure 12   A tree illustrating the Rule of Product.

whenever A and B are disjoint; while the rule of product asserts that

$$|A \times B| = |A| * |B|.$$

5 The Rules of Sum and Product (Generalized). *If an event X_1 can happen in m_1 different ways, X_2 in m_2 different ways, ..., X_k in m_k different ways, then if each pair of distinct X_i are mutually exclusive,*

(a) *the event $(X_1$ or X_2 or $\cdots$ or $X_k)$ can happen in $(m_1 + m_2 + \cdots + m_k)$ different ways;*

(b) *the event $(X_1$ and X_2 and $\cdots$ and $X_k)$ can happen in $(m_1 * m_2 * \cdots * m_k)$ different ways.*

6 Example. Consider the collection of all non-negative integers less than 10,000 that contain the digit 1. We wish to determine the percentage of all non-negative integers less than 10,000 this collection represents.

We can solve this problem by first counting all integers between 0 and 10,000 (inclusive) that do *not* contain the digit 1. We proceed by counting all strings of 4 symbols from the set $\{0, 2, 3, 4, 5, 6, 7, 8, 9\}$. There are 9 choices for the first digit, 9 choices for the second digit, 9 choices for the third digit, and 9 choices for the fourth digit. (Of course, we don't usually include leading 0's when we write a number. For this example, however, we identify 0042 as the four digit string for the number 42.) This situation can be described by the tree shown in Figure 13.

By the Rule of Product, there are $9^4 = 6561$ such numbers. This is also the number of distinct paths from the "root," the left-most node, in the tree of Figure 13 to one of the right-most nodes or "leaves." Hence, the number of (decimal) integers between 0 and 10,000 that mentions the digit 1 is 10,000 $- 6541 = 3439$. That is, slightly more than $\frac{1}{3}$ of all integers between 0 and 10,000 contain the digit 1.

We can take this result one step further, and draw from it a conclusion about "the probability of the digit 1 occurring in decimal numbers." Suppose, for example, that we fill a (rather large) bag with 10,000 numbered marbles. If we draw a marble at random and record its number, then toss it back and shake up the bag, and if we repeat this process as often as we wish, then on the average, about $\frac{1}{3}$ of the recorded decimal numbers should contain a 1.

The Rule of Product provides us with an alternative way of looking at the following result (already proved in Corollary **1.3.10**).

7 Theorem. *Any set of n elements has 2^n different subsets.*

PROOF. Let the elements of the set S be $a_1, a_2, \ldots, a_n$. Any particular subset of this set either contains the element a_1 or it does not. This means that the event $a_1 \in S$ can happen in two ways. Similarly, each of the events $a_2 \in S$, $\ldots, a_n \in S$, can happen in two ways. The number of ways in which all these

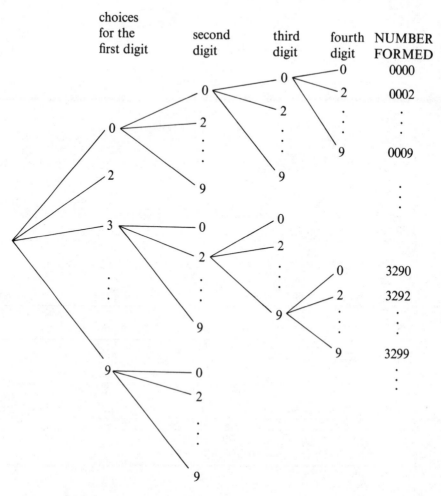

Figure 13   The tree whose paths represent 4-digit numbers not including the digit 1.

events can happen simultaneously is, by the Rule of Product, $(2 * 2 * \cdots * 2)$ n times $= 2^n$ — and this is precisely the number of distinct subsets of the set $\{a_1, a_2, \ldots, a_n\}$, as in Figure 14.   ☐

8 Example. Determine the number of ways in which a nonempty string of symbols from the set $\{A, B, C, D\}$ can be selected such that the string mentions each symbol at most once.

Four different kinds of selections are possible: The string in question may be of length 1, *or* of length 2, *or* of length 3, *or* of length 4. Let us denote these four subevents by X_1, X_2, X_3, and X_4, and the numbers of ways in which each can occur by m_1, m_2, m_3, and m_4, respectively. By the Rule of

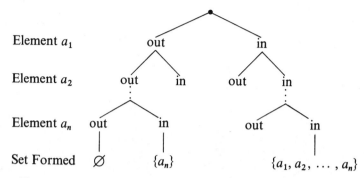

Figure 14   A tree whose paths represent the subsets of $\{a_1, \ldots, a_n\}$.

Sum (X_1 or X_2 or X_3 or X_4) can happen in $m_1 + m_2 + m_3 + m_4$ different ways.

A string of length 1, i.e., X_1, can happen in four different ways, hence $m_1 = 4$. For X_2, we have to count the number of ways in which two different symbols from $\{A, B, C, D\}$ can be written: X_2 consists of a first symbol (which can happen in four different ways), *and* a second symbol (which can happen in three different ways, the letter placed in first position *not* being a possibility for the second position). Hence, by the Rule of Product, $m_2 = 4 * 3 = 12$.

Using the Rule of Product again, and by a similar reasoning, we have $m_3 = 4 * 3 * 2 = 24$ and $m_4 = 4 * 3 * 2 * 1 = 24$. Hence, strings of length ≤ 4 with nonrepeating symbols from the set A, B, C, D can be chosen in $4 + 12 + 24 + 24 = 64$ different ways.

Let us now examine another counting technique which, although not adding anything new to the Rule of Sum and Rule of Product (Exercise 14), should be identified separately given its importance. This is the so-called "pigeonhole principle," which we first state in its simplest form.

9 The Pigeonhole Principle (simplified). *If* $(n + 1)$ *pigeons occupy* n *pigeonholes, then at least one hole will house at least two pigeons.*

10 Example. In a group of 367 people, at least 2 will have the same birthdate.

Although it is deceptively simple to state, the Pigeonhole Principle, in the form given above, is a powerful tool. We illustrate its power with the following result from Number Theory.

11 Theorem. *If* $n + 1$ *numbers are selected from the set* $\{1, 2, 3, 4, \ldots, 2n - 1, 2n\}$ *then one of the selected numbers will divide some other evenly.*

PROOF   We can write each of the $n + 1$ selected numbers as a power of 2 times an odd part. For instance, $1 = 2^0 \cdot 1$, $14 = 2^1 \cdot 7$, $40 = 2^3 \cdot 5$, etc.

Now there are exactly n odd and n even numbers between 1 and $2n$, and we shall use the n odd numbers as our pigeonholes. This is done by associating each chosen number with the odd part of its even-odd factorization (e.g., 40 goes with 5). Since $n + 1$ numbers are selected, and each is associated with one of at most n odd numbers, some odd number gets two associates. If the odd number is k then its associated pair, $2^i \cdot k$ and $2^j \cdot k$, has the desired property. $\square$

A few comments are in order concerning the preceding theorem. First of all, we classify it as a *best possible result*, in the sense that if we weaken the hypothesis in the most minimal way the theorem is no longer true. Indeed, if we select only n (instead of $n + 1$) numbers from 1 to $2n$, we do not necessarily have a pair among them that divide evenly. For example, none of the n numbers

$$\{n + 1, n + 2, n + 3, \ldots, 2n\}$$

can divide evenly any of the others in the set.

Further, the preceding theorem is called an *existence theorem*, because its statement is of the following form:

"For every configuration with properties ..., there exists an element (exist elements) in this configuration satisfying conditions ...". And the proof of the theorem is called (naturally enough!) an *existence proof*.

Existence proofs may be *constructive* or *non-constructive*. A constructive existence proof demonstrates the existence of elements satisfying the conclusion of the theorem by actually "constructing" them; e.g., by describing a step-wise method of calculating these elements. A non-constructive existence proof, on the other hand, does not exhibit such elements.

We may then call the preceding application of the pigeonhole principle a "non-constructive existence proof," since it proves the existence of a certain pair of numbers without actually calculating them. But is it really non-constructive? While this is apparently the case, the proof may nevertheless be converted into an algorithm to extract the numbers in question. For example, we can list the $(n + 1)$ selected numbers from the set $\{1, 2, 3, \ldots, 2n\}$, factor each into a power of 2 times an odd part, and finally search exhaustively for two among them that have the same odd part.

We discuss other proof techniques in Section 4.2.

Just as we generalized the Rule of Sum and Rule of Product, we can also generalize the Pigeonhole Principle. Thus, if $2n + 1$ pigeons fit into n holes, then at least one hole will house at least 3 pigeons. And if there are $(3n + 1)$ pigeons, then at least one hole will house at least 4 pigeons. For the general statement of the Pigeonhole Principle, given in the next theorem, we need to introduce new notation.

Given a non-negative number x, the *floor* of x, denoted $\lfloor x \rfloor$, is the unique integer n such that $(x - 1) < n \le x$. In other words, $\lfloor x \rfloor$ is obtained from x by throwing away its fractional part. The *ceiling* of x, denoted by $\lceil x \rceil$, is the unique integer n such that $x \leq n < (x + 1)$. Thus $\lceil x \rceil$ is the smallest integer $\geq x$. For instance, given $\pi = 3.14\ldots$, $\lfloor \pi \rfloor = 3$ and $\lceil \pi \rceil = 4$.

12 Theorem (The Pigeonhole Principle). *If m pigeons occupy n holes, $m, n \geq 1$, then one hole contains at least $\lfloor (m - 1)/n \rfloor + 1$ pigeons.*

PROOF. The largest number k such that $k * n$ is less than m is precisely $\lfloor (m - 1)/n \rfloor$. If we had exactly $n * \lfloor (m - 1)/n \rfloor$ pigeons, we could put $\lfloor (m - 1)/n \rfloor$ in each hole. But since there are m pigeons, and $n * \lfloor (m - 1)/n \rfloor$ is strictly less than m, one hole must contain more than $\lfloor (m - 1)/n \rfloor$ pigeons. $\square$

Using the general Pigeonhole Principle, we can say, for example, that in a group of 50 people, at least $\lfloor (50 - 1)/12 \rfloor + 1 = 4 + 1 = 5$ of them will be born in the same month. A more interesting illustration follows.

13 Example. We want to prove that in a group of six people, there are either three mutual friends (i.e., each pair already know each other), or else three mutual strangers (i.e., no pair has previously met).

Call the six people One, Two, ..., Six. If we consider One, the remaining five persons can be divided into two sets:

$$F = \{\text{the friends of One}\}, \text{ and}$$

$$S = \{\text{the strangers to One}\}.$$

Since we are placing five people into two groups, the Pigeonhole Principle tells us that one of the two groups has at least $\lfloor (5 - 1)/2 \rfloor + 1 = 3$ members.

Now if F has three members, then either they are three mutual strangers or at least two of them are friends. If F contains a pair of friends, then grouping them with One gives us a set of three mutual friends.

On the other hand, if it is S that has three members, then either *they* are mutual friends (in which case we are done), or else at least two of them are mutual strangers. And if S contains a pair of mutual strangers then grouping them with One gives us a set of three mutual strangers.

In all cases we have either three mutual friends or else three mutual strangers.

The situation of the preceding example can be nicely described by a diagram. We represent the six people by dots, one dot for each person, and between any two of these dots there is either a solid line (if the corresponding two people are friends) or a dotted line (if the corresponding two people are strangers). A possible situation is thus described in Figure 15.

What we have proved is that in any such diagram we find either a solid-line triangle or else a dashed line triangle. For instance, in Figure 15, {Three, Four, Six} is a solid line triangle, and {Two, Three, Five} is a dashed-line triangle.

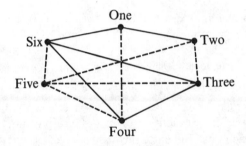

Figure 15   Solid lines link friends; dotted lines link strangers. This group does include a subgroup of three mutual friends and a subgroup of mutual strangers.

This is again a case of a "best possible" result, in the sense that six is the smallest number guaranteeing that among them we shall always find a matching threesome of mutual friends or of mutual strangers. Indeed, with five people only, we can draw Figure 16, which contains neither a solid-line triangle nor a dashed-line triangle.

The preceding discussion is a special case of a more general situation extensively studied in Combinatorial Theory, where an important and deep generalization of the Pigeonhole Principle (called Ramsey's Theorem) is studied.

PERMUTATIONS AND COMBINATIONS

Given a collection of n distinct objects, in how many different ways can we arrange them in a row? Each one of these arrangements is called a *permutation* of the n objects. For example, given objects a, b, and c, the following 6 permutations are possible:

$$abc, \; acb, \; bac, \; bca, \; cab, \; cba.$$

14 Theorem. *The number of permutations of* n *distinct objects is* $n! = n(n-1) \cdot \ldots \cdot 2 \cdot 1$.

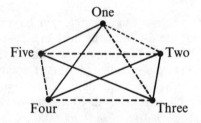

Figure 16   This group of five contains no subgroup of either three mutual friends or three mutual strangers.

PROOF. The number of ways in which we can select the first element is n. Having selected one element out of n, we are left with $(n-1)$ elements from which to choose our second element, so that, by the Rule of Product, there are $n(n-1)$ ways to select the first element, *and* the second element.

By a similar reasoning, there are $(n-2)$ ways in which we can select the third element, and $n(n-1)(n-2)$ ways to select the first, *and* the second, *and* the third elements. And so on, until we reach the nth element, which we can select in exactly one way. By the Rule of Product again, the number of ways in which we can select the first to the nth element is exactly $n\cdot(n-1)\cdot(n-2)\cdot\ldots\cdot2\cdot1 = n!$.   □

15 Example. Consider a collection of 3 objects $\{a, b, c\}$. We ask, "In how many different ways can we arrange them — not in a row — but in a circle?" This is not exactly the number of the permutations of three objects, because two distinct permutations may become identical when arranged in a circle. For instance, the three permutations abc, bca, and cab, when put in a circle,

give the same circular pattern (save for the nonessential rotation). Similarly, the three other permutations of $\{a, b, c\}$, acb, bac, and cba, give the same circular pattern. In fact, there are exactly 2 distinct ways of arranging three objects around a circle. This is a special case of the following result.

16 Theorem. *The number of arrangements of n distinct objects around a circle is $(n-1)!$.*

PROOF. We prove that there is a one-to-one correspondence, that is, a bijection between the arrangement of $\{a_1, a_2, \ldots, a_n\}$ in a circle and the permutations (i.e., arrangements in a row) of $\{a_2, a_3, \ldots, a_n\}$ — the original set of n elements from which we have deleted a_1. Since there are $(n-1)!$ permutations of the set $\{a_2, \ldots, a_n\}$, this will prove the result.

Given an arrangement of $\{a_1, \ldots, a_n\}$ in a circle, let us agree to "cut" the circle at a_1. After cutting, we "straighten" the circle, putting the next clockwise element at the left end, as in the figure

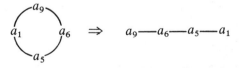

Now this mapping is clearly one-to-one: Distinct circular arrangements — arrangements that cannot be made to coincide — must lead to distinct linear

arrangements after cutting at a_1 and "straightening." The map is also onto. Any linear arrangement of $a_2, \ldots, a_n$ is the image under "cutting and straightening" of the circle formed by putting a_1 at the right end of the string, and then "bending" the string into a circle.

Thus our "circle to line" mapping is one-to-one and onto, establishing the desired bijection. □

Sometimes we do not want to count all the permutations of n objects, but only the permutations of $r \leq n$ objects out of these n objects. For example, all the permutations of 2 objects out of the 3 objects $\{a, b, c\}$ are:

$$ab, \ ac, \ ba, \ bc, \ ca, \ cb.$$

17 Notation. The number of permutations of r objects chosen from a set of n objects is denoted by $P(n, r)$. This may be read as 'n permute r', and is sometimes written ^nP_r.

We have already shown that $P(n, n) = n!$ for any positive integer n. We now want to establish the value of $P(n, r)$ for arbitrary positive integer n and non-negative integer $r \leq n$.

18 Theorem. *The number of ways of arranging in a row r objects out of a collection of n objects is given by*

$$P(n, r) = n \cdot (n - 1) \cdots (n - r + 1) = \frac{n!}{(n - r)!}$$

PROOF. This result can be proven by an argument similar to those of the two preceding theorems. We choose an alternative method here.

Note first that, for all k, $P(k, k) = k!$. This is simply a restatement of the previous theorem.

Imagine we have n marked positions which we divide into two groups, the first r positions and the remaining $(n - r)$ positions, listed in order from left to right. By the Rule of Product, the number of ways of placing n objects in these n positions is equal to the number of ways of placing r of the n objects in the first r positions multiplied by the number of ways of placing the remaining $(n - r)$ objects in the remaining $(n - r)$ positions. This means that we have

$$P(n, n) = P(n, r) \cdot P(n - r, n - r)$$

Hence, $P(n, r) = P(n, n)/P(n - r, n - r) = n!/(n - r)!$. □

19 Example. How many different 2 or 3 letter sequences are possible from the alphabet $a, \ldots, z$?

Given the 26 letters of the alphabet, we must compute $P(26, 2)$ (for the two letter sequences) and $P(26, 3)$ (for the three letter sequences). By the Rule of Sum, the desired result is $P(26, 2) + P(26, 3)$.

Now, $P(26, 2) = 26!/(26 - 2)! = 26 \times 25 = 650$ and $P(26, 3) = 26!/(26 - 3)! = 26 \times 25 \times 24 = 15,600$ — for a total of $15,600 + 650 = 16,250$ letter sequences.

There is one more combinatorial notion we want to introduce, and this is the notion of *combination*. When we select r objects out of n, and we are not interested in arranging the selected objects in any particular way, we say that we have a *combination* of r objects out of n. For instance, all the combinations of two items out of the set $\{a, b, c\}$, are:

$$\{a, b\}, \ \{a, c\}, \ \{b, c\}.$$

Since the order of the selected objects is not taken into account here, we also call a combination of r objects an *r-subset* (of a given set of n objects).

20 Notation. The number of r-subsets, or combinations of r objects, chosen from a set of n objects is denoted by $\binom{n}{r}$. This may be read as 'n choose r', and is sometimes written ^nC_r or $C(n, r)$.

21 Theorem. *The number of r-subsets of a set of n objects,*

$$\binom{n}{r} = \frac{P(n, r)}{r!} = \frac{n!}{(n - r)!r!}.$$

PROOF. Let X be an r-subset of $a_1, a_2, \ldots, a_n$, a collection of n objects. By Theorem 14, the number of ways of permuting the members of X is $r!$. Hence, by the Rule of Product, the number of ways of selecting *and* arranging r objects out of n is $\binom{n}{r} * r!$. By the last theorem, this number is also equal to $P(n, r)$. We therefore have $r! * \binom{n}{r} = P(n, r)$, proving the result. ☐

22 The Binomial Theorem. *The symbol $\binom{n}{r}$ for counting combinations occurs in the expansion of a binomial $(a + b)$ to a positive integral power n according to the formula*

$$(a + b)^n = \binom{n}{0}a^n + \binom{n}{1}a^{n-1}b + \binom{n}{2}a^{n-2}b^2 + \cdots$$

$$+ \binom{n}{n - 2}a^2b^{n-2} + \binom{n}{n - 1}ab^{n-1} + \binom{n}{n}b^n,$$

where the coefficient of the term $a^{n-r}b^r$ is $\binom{n}{r}$. The $\binom{n}{r}$ are thus also called binomial coefficients. ☐

To see why the binomial coefficients show up in the binomial theorem, consider the expression

$$\underbrace{(a + b)^n = (a + b) \cdot (a + b) \cdots (a + b)}_{n \text{ factors}}$$

and imagine how this expression might be multiplied out. Terms of the form $a^{n-r}b^r$ are formed by choosing the b summand from exactly r of the n factors in the expression. The a summand is chosen from the $(n-r)$ other factors. Now choosing b from exactly r of the n factors can happen in exactly $\binom{n}{r}$ ways, and so $\binom{n}{r}$ counts the number of ways the term $a^{n-r}b^r$ can be formed.

We conclude this section with an example that is particularly relevant to Computer Science.

23 Example. A convenient and efficient representation for the subsets of a set of elements *indexed* with integers from 1 to n, say the set

$$A = \{a_1, a_2, \ldots, a_n\}$$

is by means of a string of n *bits*, i.e., a string of n 0's and 1's. In this string of bits, if the ith bit, $1 \leq i \leq n$, is 0 then we agree that a_i is *not* a member of the subset, and if it is 1 then a_i is a member. For instance, if $n = 16$ and $A = \{a_1, a_2, \ldots, a_{16}\}$ then the string

$$1\ \ 1\ \ 0\ \ 1\ \ 1\ \ 0\ \ 0\ \ 0\ \ 0\ \ 0\ \ 0\ \ 0\ \ 0\ \ 0\ \ 0\ \ 1$$

represents the subset $\{a_1, a_2, a_4, a_5, a_{16}\}$. In fact we have here a bijective correspondence between the collection of all bit-patterns of length n and the collection of all subsets of $A = \{a_1, a_2, \ldots, a_n\}$. We immediately conclude that there are exactly 2^n bit-patterns of length n (since there are $2^{|A|}$ subsets of A), and there are exactly $\binom{n}{r}$ bit-patterns with r one's and $(n-r)$ zero's (this is the number of r-subsets of A). (The reader should recall the discussion of characteristic functions following Fact **1.3.7**.)

EXERCISES FOR SECTION 3.1

1. In how many ways can a deck of 52 cards be shuffled?

2. A committee has 8 members. In how many ways can two (disjoint) subcommittees, each of size three, be chosen?

3. Suppose 5 points are chosen from inside a square with edge length 2. Show that some pair of points must be within $\sqrt{2}$ of each other.

4. In how many ways can 6 different colored beads be strung on a necklace?

5. For $n > m > k$ which is larger,

$$\binom{2n}{n}, \quad \binom{2m}{m}, \quad \text{or} \quad \binom{2k}{k}?$$

6. Show that $n \cdot \binom{n-1}{r} = (r+1) \cdot \binom{n}{r+1}$. Can you prove your result without using the explicit formula for $\binom{n}{k}$?

7. A bit-string of 0's and 1's is said to have *even parity* if 1 occurs in the string an even number of times; otherwise it is of *odd parity*. Given all bit strings of length n, for

some positive integer n, how many of them are of even parity? And how many are of odd parity?

8. A coin is flipped n times, for some positive integer n. (a) Count the number of ways of getting exactly one head. (b) Count the number of ways of getting exactly r heads, for $0 \leq r \leq n$.

9. Use the binomial theorem to prove the identity

$$\binom{n}{0} + \binom{n}{1} + \cdots + \binom{n}{n} = 2^n.$$

10. Prove that the total number of even subsets (i.e., subsets with an even number of elements) of a set of n elements is equal to the total number of its odd subsets by first using the Binomial Theorem to prove the identity

$$\binom{n}{0} - \binom{n}{1} + \binom{n}{2} - \cdots \pm \binom{n}{n} = 0$$

where the minus and plus signs alternate and the last sign may be either $+$ or $-$ depending on whether n is even or odd.

11. Prove that the number of ways of selecting r objects from a set of n distinct objects, where order does not count and repetitions are allowed is $\binom{n+r-1}{r}$. (Hint. A typical selection of r out of n elements $a_1, a_2, \ldots, a_n$ with repetitions allowed can be represented by a string of symbols of the form

$$a_1 a_1 a_1 || a_3 | a_4 ||| a_7 a_7 a_7 a_7 || a_9 a_9$$

where there is a $|$ between the a_i's selected and the a_{i+1}'s selected. Two consecutive $|$'s means that some a_i has not been selected. Three $|$'s in a row means that two a_i's have been omitted as with a_5 and a_6. The total number of such selections of r objects is the number of strings of the above form, where there are exactly r a's and $(n-1)$ "vertical bars.")

12. (a) In how many ways can three numbers be selected from the set $\{1, 2, \ldots, 50\}$ such that their sum is even?
    (b) Repeat (a) when the sum is required to be odd.
    (Hint. The sum of three integers is odd if and only if either one or three of them are odd.)

13. (a) In how many ways can three numbers be selected from the set $\{1, 2, \ldots, 99\}$ such that their sum is a multiple of 3?
    (b) Generalize (a) to selections of three numbers from the set $\{1, 2, \ldots, 3n\}$.

14. Argue that the rule of sum (generalized) can be restated in the following form: If a finite set X of objects is divided into mutually disjoint subsets $X_1, X_2, \ldots, X_k$, then the number of objects in X can be determined by finding the number of objects in each of the sets $X_1, X_2, \ldots, X_k$, and adding. Using this form of the rule of sum, deduce the pigeonhole principle.

15. A square array of dimension $n \times n$, with all the entries consisting of the integers $1, 2, 3, \ldots, n^2$, is called a *magic square* if the sum S of the integers in each row, in each column, and in each of the two diagonals is the same. The number S is called

the *magic sum* of the magic square. For example, the following square arrays are magic squares of order 3 and 4, respectively:

6	1	8
7	5	3
2	9	4

13	2	3	16
12	7	6	9
8	11	10	5
1	14	15	4

The magic sum of the first is 15, that of the second is 34. Magic squares were studied in antiquity in China and were introduced in medieval times in Europe, where they were supposed to protect against evils.

(a) Show that there cannot exist a magic square of order 2 (i.e., dimension 2×2).

(b) Prove that all magic squares of the same order n must have the same magic sum S. (E.g., all magic squares of order 3 have the sum 15.) Find an expression for S as a function of n.

16. A square array of dimension $n \times n$ is called a *latin square* of order n if each row and column is a permutation of the numbers $1, 2, \ldots, n$. An example of a latin square of order 4 is

1	4	3	2
2	1	4	3
3	2	1	4
4	3	2	1.

Whereas "magic squares" in the previous exercise seem to be a mathematical curiosity with little practical value, latin squares have played an important role in the development of combinatorial theory, and have proved to be useful in certain areas of applied mathematics such as statistics.

(a) Find an upper bound on the number of latin squares of order n. Justify your answer. (It is considerably more difficult to find the exact number of latin squares of order n — if you can find this exact number, you probably need not take this course ...)

Two latin square A and B of the same order are said to be *orthogonal* if by juxtaposing A and B we get n^2 distinct pairs of entries. For example, if A and B are

3	2	1
2	1	3
1	3	2

2	3	1
1	2	3
3	1	2

then their justaposition results in the following array

(3, 2)	(2, 3)	(1, 1)
(2, 1)	(1, 2)	(3, 3)
(1, 3)	(3, 1)	(2, 2).

Since all entry pairs in the resulting array are distinct, A and B are orthogonal.

(b) Show that if A and B are orthogonal latin squares, then in each of A and B there are n positions which lie in n different rows *and* n different columns such that the entries in these positions are the integers $1, 2, 3, \ldots, n$ in some order.

3.2 Trees and Recurrences

BASIC DEFINITIONS AND PROPERTIES OF TREES

We have already encountered "trees" in Chapter 2, and in our discussion of counting principles in the previous section. A typical tree is displayed in Figure 17. The node A is called the root; nodes F, G, H, J, K, L, and M are called leaves. With complete disregard for the botanical facts, computer scientists represent trees pictorially with the root above the leaves.

We can see that every node is either a leaf, or has edges below it which lead to other nodes. Each node in a tree can be regarded as the root for the subtree consisting of the tree structure below it. This observation is the basis for our next definition.

1 Definition. A (rooted) *tree* T is a nonempty set of labeled nodes with one distinguished node, called the *root* of the tree; the remaining nodes are partitioned into $m \geq 0$ disjoint subtrees $T_1, T_2, \ldots, T_m$. Nodes having no subtrees are called *leaves*; the remaining nodes are called *internal* nodes.

Set theoretically, we can say that the tree of Fig. 17 is:

$T = \{A, B, C, D, E, F, G, H, I, J, K, L, M\}$

Root of T: A

Subtrees of T: $T_1 = \{B, E, K, L\}$, $T_2 = \{C, F, G\}$, $T_3 = \{D, H, I, J, M\}$

Root of T_1: B
Subtrees of T_1: $T_4 = \{E, K, L\}$

Root of T_2: C
Subtrees of T_2: $T_5 = \{F\}$, $T_6 = \{G\}$

Root of T_3: D
Subtrees of T_3: $T_7 = \{H\}$, $T_8 = \{I, M\}$, $T_9 = \{J\}$

Root of T_4: E
Subtrees of T_4: $T_{10} = \{K\}$, $T_{11} = \{L\}$

Root of T_8: I
Subtrees of T_8: $T_{12} = \{M\}$

Subtrees $T_5, T_6, T_7, T_9, T_{10}, T_{11}, T_{12}$ are all leaves.

It should be clear that the graphic representation of the tree is not only easier to grasp, but also more compact than the set-theoretic representation.

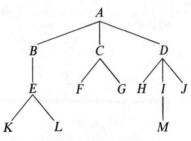

Figure 17   A tree.

We shall now very briefly examine some of the issues related to trees. Because the notation of a tree captures the idea of a hierarchy or a structured organization, it is one of the most useful notions in computer science.

Let us introduce some additional terminology that is helpful when we talk about trees. The terms used here are of mixed genealogical-botanical origins. Thus the *root A* in the tree of Figure 17 has three *branches*, each connecting it to a different subtree. Furthermore, all the nodes in a tree of a subtree are said to be *descendants* of its root; conversely, the root is an *ancestor* of all its descendants. We also refer to the root in a tree (or a subtree) as the *parent* of the roots of its subtrees; these nodes are in turn the *children* or *immediate successors* of the root.

A *path* is a succession of consecutive branches connecting the root to a leaf. In Figure 17, for example, *A D I M* is a path. In a tree there are clearly as many distinct *finite* paths as there are leaves. Note that we did not restrict a tree to be finite. A tree may have infinitely many nodes either because it has an infinite path or because some node has infinitely many branches. This fact, which is known as König's lemma, requires proof — see Exercise 14. When all the paths in a tree are finite, we can talk meaningfully about the *height* of the tree, which is the number of branches in a longest path.

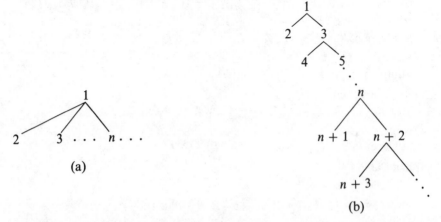

Figure 18   Two examples of infinite trees. (a) Node 1 has infinitely many branches. (b) Node 1 is the root of an infinitely descending path.

2 Example. Consider the tree in Figure 19 whose root is labeled with the finite set $\{a, b, c\}$, and its remaining nodes with subsets of $\{a, b, c\}$, such that the immediate successors of a node X are labeled with the subsets of the set associated with X obtained by deleting one element.

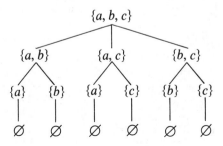

Figure 19   A tree of subsets.

This tree has six paths from the root to its leaves, all are of length 3, and therefore the tree height is also 3.

In all trees considered so far, the order in which the siblings of a node are drawn (from left to right) was not relevant. Thus

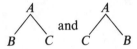

were identical trees. Often we need to talk about *ordered* trees, where the relative order of subtrees is important.

3 Example. Consider the algebraic expression $((a - 6) * b)/(c - 3)$ represented by the tree of Figure 20a.

This is an ordered tree of height 3 (the length of the longest path). We choose to have it "ordered" because once the symbols are given their usual

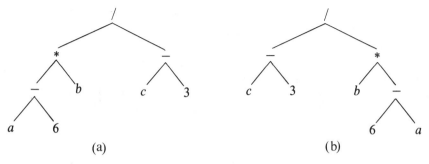

Figure 20   Two trees representing arithmetic expressions. (a) $((a - 6) * b)/(c - 3)$. (b) $(c - 3)/(b * (6 - a))$. Note that order of nodes plays a crucial role on this interpretation.

meanings, permuting subtrees changes the meaning of the tree representation. Thus, the tree in Figure 20b has the "meaning": $(c - 3)/(b * (6 - a))$ which is obviously different from the original expression in its meaning.

Just as we earlier defined trees in general in terms of sets, we can also define ordered trees in terms of ordered sets. For the present we shall not burden our presentation with such formal definition. Instead, we use the (more intuitive) drawn diagrams both for trees and ordered trees.

An important variant of the ordered trees is the class of "binary trees." Their chief characteristic is that any node can be parent of at most two children.

4 Definition. A *binary tree* T is a finite set of nodes which is either empty or consists of a root and two binary subtrees T_L and T_R in that order, called the *left* and *right* subtrees of T.

It is important to note that binary trees are *not* a subclass of ordered trees. Indeed, a tree or an ordered tree cannot be empty, while a binary tree can. Furthermore, the way we draw the subtrees of a root in binary trees is important; for instance,

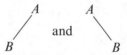

are different as binary trees because the left tree has an empty right subtree, while the right tree has an empty left subtree. As trees and ordered trees, on the other hand, both are identical to

While a node in an ordered tree may be thought of as a root with 0 or more subtrees (ordered from left to right), nodes in binary trees always have exactly two subtrees (one or both of which being possibly empty).

A *forest* of trees (or ordered trees, or binary trees) is a collection of the given type. If we list the members of a forest in some prescribed order, then we have an *ordered forest*.

5 Example. The tree in Figure 20a was given as an ordered tree. We can also look at it as a binary tree. If we remove its root node, /, we get a forest made up of the two subtrees of the original binary tree — namely,

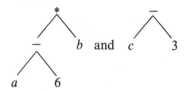

In fact, if we give the node labels their usual meanings, we should view the forest as an ordered forest, because if we connect them back by taking them again as subtrees of a binary tree it will make a difference which of the trees is on the left.

We now state and prove an important counting property of binary trees. This property was implicit in several parts of the preceding section, where we talked about counting principles, permutations, and combinations. To establish this result, we need the notion of "level" in a tree, which we define recursively. First, the *level* of the root node is 0. Second, if a node is at *level n*, then its children are at level $(n + 1)$. This means that the height of a tree, which we defined earlier, is also the maximum level of any node in the tree.

6 Proposition. *The number of nodes at level i in a binary tree is $\leq 2^i$.*

PROOF. By induction on i, if $i = 0$, the binary tree has exactly one node at this level: the root node; and the total number of nodes is indeed $2^0 = 1$.

Assume that the result is true for an arbitrary integer i. We want to prove it for $(i + 1)$. Consider a binary tree T which has k nodes at its ith level. By the induction hypothesis, $k \leq 2^i$. Each of these k nodes can have at most two children, so that the number of nodes at level $i + 1$ is at most $2 \cdot k \leq 2 \cdot 2^i = 2^{i+1}$, as desired. □

7 Corollary. *The number of nodes in a binary tree of height n is $\leq (2^{n+1} - 1)$.*

PROOF. By the preceding result the maximum number of nodes at level i is 2^i. If every level has as many nodes as possible, and the tree height is n, then the total number of nodes is bounded by

$$\sum_{0 \leq i \leq n} 2^i = 2^0 + 2^1 + \cdots + 2^n.$$

By induction on n, we shall prove that this sum is equal to $(2^{n+1} - 1)$.

Basis Step. For $n = 0$ we clearly have that $2^0 = 2^{0+1} - 1 = 1$.
Induction Step. For $n \geq 1$, suppose that $2^0 + 2^1 + \cdots + 2^{n-1} = 2^n - 1$, and suppose we double both sides: $2(2^0 + 2^1 + \cdots + 2^{n-1}) = 2(2^n - 1)$. Then $2^1 + 2^2 + \cdots + 2^n = (2^{n+1} - 2)$, and so $2^0 + 2^1 + 2^2 + \cdots + 2^n = 2^{n+1} - 1$. □

level 0

level 1

level 2

level 3

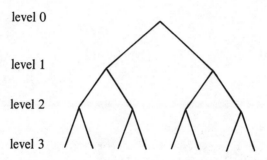

Figure 21   A full binary tree of height 3.

8 Definition. A *full binary tree* of height n is a binary tree such that each of its nodes has two children, except that those at level n are all leaves.

A full binary tree of height 3 is shown in Figure 21.

In a full binary tree of height n, every level i, $0 \leq i \leq n$, contains exactly 2^i nodes, and the total number of nodes is $(2^{n+1} - 1)$. Thus the bounds obtained in Corollary 7 above are in fact attained by full binary trees.

9 Definition. We say a binary tree of height n is *complete* if at every level i, $0 \leq i \leq (n - 1)$, it has exactly 2^i nodes, *and* if all nodes at level $(n - 1)$ with two children are to the left of those nodes with no children, *and* there are no nodes with only one child, as in Figure 22.

10 Corollary. *The height of a binary tree with n nodes, for $n \geq 1$, is at least* $\lfloor \log_2 n \rfloor$.

PROOF. To minimize the height of a tree built from n nodes, fill all levels up to say level h, and then distribute the remaining nodes at level h. Then levels 0 up to $(h - 1)$ contain exactly $(2^h - 1)$ nodes, while level h contains at least one node and no more than 2^h nodes. Hence, the minimum number of nodes in a complete binary tree of height h is

$$(2^h - 1) + 1 = 2^h,$$

and the maximum number of nodes in such a tree is

$$(2^h - 1) + 2^h = 2^{h+1} - 1.$$

Hence $2^h \leq n \leq 2^{h+1} - 1$. If $n = 2^h$ we have $h = \log_2 n$, and if $n = 2^{h+1} - 1$ we have $h + 1 = \log_2(n + 1)$. It is left as an exercise to show that in the latter case $h = \lfloor \log_2 n \rfloor$. Thus, in both cases we have $h = \lfloor \log_2 n \rfloor$.   □

Counting the nodes present in various kinds of trees, and especially in binary trees, is a fundamental aspect of a topic in Computer Science called "Algorithm Analysis." We shall give an important example of algorithm analysis in Section 3.3.

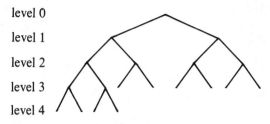

level 0

level 1

level 2

level 3

level 4

Figure 22   A complete binary tree of height 4.

TRAVERSALS OF BINARY TREES

In many applications, we need to "traverse" all the nodes in a tree or a
forest; that is, we wish to visit all the nodes in a tree or forest in some systematic manner. We briefly discuss here some of these traversal procedures for
binary trees. We consider the so-called "preorder," "inorder," and "postorder" procedures, which we define recursively.

11 Definition. Given a binary tree T, and its two subtrees T_L and T_R, we can
traverse the nodes of T in:

Preorder: 1. Visit the root of T;
                2. Traverse the left subtree T_L in preorder;
                3. Traverse the right subtree T_R in preorder.
Inorder: 1. Traverse the left subtree T_L in inorder;
                2. Visit the root of T;
                3. Traverse the right subtree T_R in inorder.
Postorder: 1. Traverse the left subtree T_L in postorder;
                2. Traverse the right subtree T_R in postorder;
                3. Visit the root of T.

The origins of the terms "preorder," "inorder," and "postorder," will
become clear as some examples are worked out.

12 Example. Consider the binary tree of Figure 20a, representing the
expression $((a - 6) * b)/(c - 3)$. Figure 23 gives integer labels that indicate
the order in which nodes are visited according to each of the traversal
procedures.

If we list the original nodes' labels in Figure 20a according to each of the
traversals, we get:

$$\text{Preorder:} \quad / * - a\, 6\, b - c\, 3$$

$$\text{Inorder:} \quad a - 6 * b / c - 3$$

$$\text{Postorder:} \quad a\, 6 - b * c\, 3 - /$$

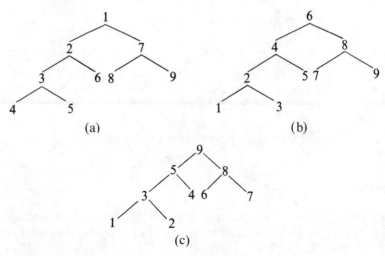

(a)   (b)

(c)

Figure 23   Illustrating three ordering of nodes of a tree: (a) Preorder. (b) Inorder. (c) Postorder.

Note that the original algebraic expression $((a - 6) * b)/(c - 3)$ is identical to the inorder listing modulo the insertion of parentheses. Here the parentheses are necessary in order to distinguish between two different expressions with the same sequence of symbols (save for the parentheses). For instance, the expression $(a - (6 * b))/(c - 3)$ has a binary tree representation of Figure 24. While this tree clearly represents an algebraic expression different from the preceding binary tree, its inorder listing is, nevertheless, the same:

$$\text{Inorder:} \quad a - 6 * b / c - 3.$$

We therefore conclude that it is not always possible to reconstruct a binary tree uniquely from an inorder presentation of its nodes. In this sense, inorder listings are ambiguous.

The beauty of preorder and postorder expressions is that they *can* be used to reconstruct uniquely corresponding binary trees. We leave it to the reader to check that the preorder and postorder listings above uniquely describe the binary tree from which we have derived them. We thus say that preorder and postorder expressions are *unambiguous*. As a result, we do not

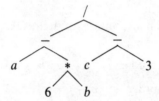

Figure 24   Binary tree representing $(a - (6 * b))/(c - 3)$.

need to insert parentheses in preorder and postorder expressions to remove their ambiguity.

We shall look at preorder from another perspective in Section 5.3.

RECURRENCE RELATIONS

We have now seen that counting methods, and in particular counting methods on trees, are intimately related to inductive proofs and recursive definitions. There is still another important notion which is closely related to counting principles considered so far: this is the notion of *recurrence* or *recurrence relations*.

In a way we have already encountered this notion. A recursive definition always establishes a recurrence relation, and conversely, every recurrence relation can be construed as a recursive definition. However, the use of a recurrence relation (or *difference equation*) in mathematics usually calls for an explicit closed-form solution, while the use of a recursive definition in computer science does not.

13 Example. Consider, the following system of equations involving the function $f: \mathbf{N} \to \mathbf{N}$

$$(1) \quad f(0) = 1;$$
$$(2) \quad f(n) = n * f(n - 1), \text{ for } n \geq 1.$$

This is clearly the "recursive definition" of the factorial function. But if we want to find a closed-form expression for a function satisfying the above equations, we can just as well view (2) as a "recurrence relation" with (1) as its "initial condition." Admittedly the distinction between "recurrence relation" and "recursive definition" in this case seems contrived — and it is.

14 The Fibonacci Sequence. A situation that perhaps deserves more the name "recurrence relation" is the following one. Let $\{a_0, a_1, a_2, \ldots\}$ be an infinite sequence of integers satisfying the following conditions:

$$(1) \quad a_0 = 1,$$
$$(2) \quad a_1 = 1,$$
$$(3) \quad a_n = a_{n-1} + a_{n-2}, \text{ for } n \geq 2.$$

The third equation relates an arbitrary number a_n in the sequence, for $n \geq 2$, to the two numbers immediately preceding it, a_{n-1} and a_{n-2}. And the values specified by (1) and (2) are "initial conditions" which are required before we can start assigning values to $a_2, a_3, a_4, \ldots$, in succession using relation (3). It is clear how we can generate the numbers in the above sequence:

a_0	a_1	a_2	a_3	a_4	a_5	a_6	a_7	a_8	a_9	$\cdots$
1	1	2	3	5	8	13	21	34	55	$\cdots$

This sequence of numbers is called the Fibonacci sequence. It was first studied by Leonardo of Pisa, alias Fibonacci, around the year 1200 in his work *Liber Abaci*. In it, he poses the so-called "rabbit problem" according to which rabbits reproduce in the following manner: every pair of rabbits (at least one month old) will produce one pair of rabbits as offspring every month. To simplify the analysis, it is assumed that rabbits never die and never stop reproducing. How many rabbits are there after n months?

In the Fibonacci sequence $\{a_0, a_1, a_2, a_3, \ldots\}$, a_n is the number of rabbit pairs in the population after n months. If we denote by b_n the number of pairs born during the nth month, and by c_n the number of at least one month old pairs at the end of the nth month, we have $a_n = b_n + c_n$. According to the breeding law mentioned above:

$$c_{n+1} = b_n + c_n = a_n, \quad \text{i.e., the old population at time } n+1 \text{ is the entire population at time } n; \text{ and}$$

$$b_{n+1} = c_n \qquad \text{i.e., every old pair at month } n \text{ produces a newborn pair during month } n+1.$$

In month $n + 2$ the same events are repeated:

$$c_{n+2} = b_{n+1} + c_{n+1} = a_{n+1},$$
$$b_{n+2} = c_{n+1}.$$

Combining these equations we obtain:

$$a_{n+2} = b_{n+2} + c_{n+2} = c_{n+1} + a_{n+1} = a_n + a_{n+1},$$

that is,

$$a_{n+2} = a_{n+1} + a_n$$

which is identical to recursive relation (3) stated earlier.

It should be added that while the Fibonacci sequence is of little use to breeders, it is of great importance in mathematics and computer science (in particular, algorithm analysis).

We can reformulate the Fibonacci problem in terms of a function

$f : \mathbf{N} \to \mathbf{N}$. Recurrence relation (3) and its initial conditions (1) and (2) then take the form

(1') $f(0) = 1$,

(2') $f(1) = 1$,

(3') $f(n) = f(n - 1) + f(n - 2)$, for $n \geq 2$.

We can call this function f the *Fibonacci function*. The correspondence between the Fibonacci sequence and the Fibonacci function is obvious: for all $n \in \mathbf{N}$, a_n is the same as $f(n)$.

Now, finding a closed-form expression for the Fibonacci function demands more effort than was required in the factorial example. There are well-studied techniques to find solutions of recurrence relations, which we shall not cover here. Let us at least mention the solution that can be found using some of these techniques for the Fibonacci function:

$$f(n) = \frac{1}{\sqrt{5}} \left(\frac{1 + \sqrt{5}}{2} \right)^{n+1} - \frac{1}{\sqrt{5}} \left(\frac{1 - \sqrt{5}}{2} \right)^{n+1}.$$

This is *not* an expression that can be guessed by looking long enough at the above recurrence relation. Of course, once the formula is found, it can be verified by induction (Exercise 6).

Our next example of a recurrence is one of the most important in mathematics.

15 Example. Earlier in this chapter, in **3.1.22**, we introduced the binomial coefficients, which we denoted $\binom{n}{r}$. Binomial coefficients can be given one of two meanings (at least), which we shall call "combinatorial" and "factorial":

(1)   *combinatorial* — we define $\binom{n}{r}$ to be the number of ways in which we can select $r \leq n$ objects from a set of n objects;

(2)   *factorial* — we define $\binom{n}{r}$ to be $n!/(n - r)!r!$.

Either one of the two meanings, (1) and (2), can be chosen as the basic definition of $\binom{n}{r}$, and we can then derive the other meaning.

Here we want to show that $\binom{n}{r}$ satisfies a certain recurrence relation; namely:

$$\binom{n}{r} = \begin{cases} 1, & \text{if } r = 0; \\ 1, & \text{if } r = n; \\ \binom{n-1}{r-1} + \binom{n-1}{r}, & \text{if } 0 < r < n. \end{cases}$$

We argue, using the combinatorial meaning of $\binom{n}{r}$, that binomial coefficients indeed satisfy the above relation.

Concerning the "initial conditions" $\binom{n}{0}$ and $\binom{n}{n}$, it is evident that there is exactly one way of selecting 0 things and n things, respectively from a set of n things. Hence $\binom{n}{0} = \binom{n}{n} = 1$.

Consider now the case when $0 < r < n$. If we select a subset of r objects from the set, say $\{a_1, a_2, \ldots, a_n\}$, then we can do it in one of two ways:

$$\textit{case A} \text{ — one of the } r \text{ objects is } a_1;$$

$$\textit{case B} \text{ — none of the } r \text{ objects is } a_1.$$

In case A we then choose the rest of the subset ($r - 1$ objects) from the remaining $n - 1$ objects $\{a_2, a_3, \ldots, a_n\}$. Thus A can happen in $\binom{n-1}{r-1}$ ways. In case B we have to select all r objects from the remaining $n - 1$ objects $\{a_2, a_3, \ldots, a_n\}$. Hence B can happen in $\binom{n-1}{r}$ ways. By the Rule of Sum, we can select r things out of the n objects $\{a_1, a_2, a_3, \ldots, a_n\}$ in

$$\binom{n-1}{r-1} + \binom{n-1}{r}$$

ways. We have thus proved the desired recurrence relation.

We could have also started from the factorial meaning, i.e., from the fact that $\binom{n}{r} = n!/(n-r)!r!$, and again showed that binomial coefficients satisfy the above recurrence relation. In either case we start from an already given definition — here either the combinatorial or factorial definition of $\binom{n}{r}$ — and then prove that it satisfied a certain recurrence relation (which is *not* readily inferred from this definition).

One final point. Given the sequence of values generated by a recurrence relation, there is a natural way of calculating these values using trees. Suppose for instance that we want to compute the 6th number in the Fibonacci sequence, $f(5)$. Starting with $f(5)$, two values $f(4)$ and $f(3)$ are required before we can compute it; and to compute $f(4)$ and $f(3)$, we first need to know the values of $f(3)$ and $f(2)$, and of $f(2)$ and $f(1)$, respectively.

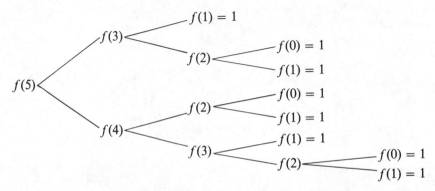

Figure 25   A parsing tree for the Fibonacci number, $f(5)$, corresponding to $f(n) = f(n - 1) + f(n - 2); f(0) = f(1) = 1$

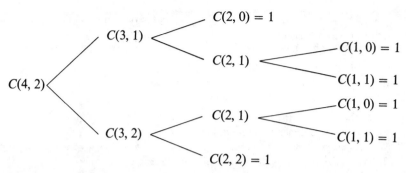

Figure 26   A parsing tree for the binomial coefficient $C(4, 2)$ corresponding to $C(n, r) = C(n - 1, r - 1) + C(n - 1, r); C(n, 0) = C(n, n) = 1.$

We can proceed in this fashion "backward" until we reach $f(0)$ and $f(1)$, whose values are given directly by the initial conditions. This inductive process is best described by the tree of Figure 25. To find the value of $f(5)$ we have to climb up to the root $f(5)$ from the leaves, using the fact that the value of a parent node is the sum of the values of its two children.

A similar evaluation of the binomial coefficient $\binom{4}{2}$ is given by the tree of Figure 26.

EXERCISES FOR SECTION 3.2:

1. Given the ordered tree T below:

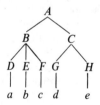

   (i)   What is the root of the tree?
   (ii)   What nodes are the ancestors of d?
   (iii) How many leaves are there?
   (iv)   How many internal nodes are there?

2. What expression tree corresponds to the postfix expression

$$5\,7\,9 + *\,2\,1 + 1 * ?$$

3. List all the expression trees according to the infix expression

$$2 + 3 * 4 + 5.$$

4. Show that if $h + 1 = \lceil \log_2 (n + 1) \rceil$, where h and n are positive integers, then $h = \lfloor \log_2 n \rfloor$ (cf. Corollary 10).

5. Let $a_0, a_1, a_2, \ldots$ be the Fibonacci sequence. Prove the following identities in two different ways: first, inductively (easy) and, second, by appropriate grouping of terms (more difficult).
   (a) $a_0 + a_1 + a_2 + \cdots + a_{n-1} = a_{n+1} - 1$
   (b) $a_0 + a_2 + a_4 + \cdots + a_{2n} = a_{2n+1}$
   (c) $a_0^2 + a_1^2 + a_2^2 + \cdots + a_n^2 = a_n a_{n+1}$

6. Verify the formula $f(n) = (1/\sqrt{5})[((1 + \sqrt{5})/2)^{n+1} - ((1 - \sqrt{5})/2)^{n+1}]$ for the Fibonacci function, using induction on n.

7. We define inductively a function f: {binary trees} $\rightarrow N$. Recall that we denote by T_L and T_R the left and right subtrees of a binary tree T.

$$f(T) = \begin{cases} 0 & \text{if } T \text{ is empty;} \\ \max(f(T_L), f(T_R)) & \text{if } f(T_L) \neq f(T_R); \\ f(T_R) + 1 & \text{if } f(T_L) = f(T_R). \end{cases}$$

Such a function f is called a *Strahler numbering* of binary trees.
   (a) What is the Strahler number of a full binary tree of height n? And the Strahler number of a complete binary tree of height n? Give a general relationship between the height of a binary tree and its Strahler number.
   (b) Devise an inductive procedure based on the postorder traversal to compute the Strahler number of a binary tree.

8. Prove that if a tree has n nodes, then it has exactly $(n - 1)$ edges.

9. Prove that the number of internal nodes in a binary tree of height $h > 0$ is less than $2^h - 1$.

10. Let T be a binary tree where every node has either no children or exactly two children, i.e., we preclude the possibility that a node may have one child. Let T have n leaves, for some integer $n \geq 1$, and $\ell_1, \ell_2, \ldots, \ell_n$ be the respective levels of these n leaves. Let $C(i) = $ the number of leaves at level i.
   (a) Show that $\sum_{1 \leq i \leq n} C(i) \cdot (1/2^i) = 1$. (Hint: It is helpful to first think in terms of full binary trees, then in terms of complete binary trees, and finally in terms of binary trees in general — all satisfying the condition that a node cannot have exactly one child node.)
   (b) Prove that $\max(\ell_1, \ell_2, \ldots, \ell_n) \geq \log n$.

11. In the present exercise all trees are ordered binary trees where all internal nodes have exactly two children. If such an ordered binary tree has exactly n leaves, we say that its *order* is n. We shall label the leaves of a tree of order n with the letters $x_1, x_2, \ldots, x_n$, from left to right; further, if two nodes with labels u_1 and u_2 are the left and right children of the same node, we label the latter with $(u_1 u_2)$ — parentheses included. For example, a tree of order 6 with the labeling just described looks as shown in Figure 27.

   (a) Argue that there is a bijective correspondence between the collection of all such trees of order n and the different ways we can parenthesize the product $x_1 x_2 \cdots x_n$. (To parenthesize a product means to insert enough parentheses so that every subproduct has exactly two factors — with the convention that if $n = 1$ we have one factor x_1 and do not need to parenthesize, and if $n = 2$ we have two factors $x_1 x_2$ and only one possible parenthesization $(x_1 x_2)$.)

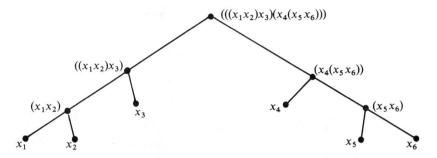

Figure 27   A tree with nodes labeled by parenthesized products.

(b) Prove that the total number of left-going (or, right-going) edges in such a tree of order n is $(n - 1)$. Conclude that the number of parentheses required to parenthesize n factors is $(n - 1)$ left parentheses and $(n - 1)$ right parentheses.

(c) If $f: \{1, 2, \ldots\} \to \{1, 2, \ldots\}$ is the function that counts the number of different ways a product of $n \geq 1$ factors can be parenthesized, show that f satisfies the following recurrence relation and initial condition:

(1) $f(1) = 1$;

(2) $f(n) = f(1)f(n - 1) + f(2)f(n - 2) + \cdots + f(n - 1)f(1)$
     $= \sum_{1 \leq i \leq n-1} f(i)f(n - i)$, for $n \geq 2$.

(The solution of this recurrence relation, by techniques not examined in this book, is $f(n) = (1/n)\binom{2n-2}{n-1}$.)

12. Give preorder and postorder representations of the tree given in Figure 24.

13. Show that for fixed n and $0 \leq k \leq 2n$, the expression $\binom{2n}{k}$ is largest when $k = n$.

14. Prove König's lemma: "If a finitely branching tree has infinitely many nodes, then it has an infinite branch." (Hint: use the following extension of the pigeonhole principle: "If infinitely many pigeons are placed in finitely many holes, then some hole holds infinitely many pigeons." To use this principle in the proof of König's lemma, consider the finitely many immediate descendants of the root node. Collectively, they cover the infinitely many nodes (pigeons) present in the tree with the root omitted. So one of these descendants must cover infinitely many nodes. Put that node (along with the root) in the to-be-constructed infinite branch. The chosen node is itself the root of a finitely branching subtree with infinitely many nodes, so the same argument may be repeated.)

15. What function f solves the recurrence

$$f(0) = 1$$

$$f(n) = 2 * f(n - 1)?$$

16. Let the function $h(n)$, for every integer $n \geq 0$, be

$$h(n) = 1^3 + 2^3 + \cdots + (n - 1)^3 + n^3.$$

Verify a recurrence relation with an appropriate initial condition for which $h(n)$ is the solution.

$$(\text{Answer:} \quad h(0) = 0$$

$$h(n) = h(n-1) + n^3.)$$

17. Consider a $1 \times n$ chessboard. Suppose we color each square of the chessboard black or white. For $n \in \{1, 2, 3, \ldots\}$, define the function $h(n)$ to be the number of colorings in which no two adjacent squares are colored black — although two adjacent squares may be colored white. Find a recurrence relation satisfied by $h(n)$, then find a closed-form formula for $h(n)$.

18. Find a closed-form expression for the function $h(n)$, which satisfies the following recurrence relation:
   (1) $h(n) = 4 * h(n-2)$, for $n = 2, 3, 4, \ldots$
   (2) $h(0) = 0$
   (3) $h(1) = 1$

3.3 An Example of Algorithm Analysis

The "analysis of algorithms" is an important subject for both practical and theoretical computer science. In this section we illustrate the nature of this topic by means of a single example: the problem of searching an array (or a file) for an item x. Our illustration of analysis techniques will be based on the theory laid down earlier in the chapter.

Admittedly we have not yet explained what the "analysis of algorithms" actually involves — and we shall not try to discuss this concept in general terms. Instead we shall show how to analyze algorithms for array searching, and use this to give the reader a taste for the "style" of the subject. The general question is this — given different algorithms to handle a given task, how can we compare their "efficiency" or "complexity"?

SEQUENTIAL VERSUS BINARY SEARCH

Consider the problem of searching for a particular record in some list. For our purposes we can simplify the problem in the following way. Every record in question is associated with a *search key*, a natural number which identifies it uniquely, and all such search keys are stored sequentially in a one-dimensional array A. If we have $n \geq 0$ records, then we have n search keys stored in array cells $A[1]$, $A[2]$, ..., $A[n]$. The problem of accessing a particular record with search key x is therefore the problem of finding an index value i, $1 \leq i \leq n$, such that $A[i] = x$.

The simplest procedure to solve this problem is to *sequentially search* the entries of the array A, in the order in which they are stored from 1 to n, until one is found to equal x. This algorithm is described by the flow-chart of Figure 28.

Let us agree to measure the cost of a search by counting the number of entries that have to be examined in the array A. Clearly, the smallest possible number of look-ups is 1, which occurs when $A[1]$ happens to store the value x; and the largest such number is n, which occurs when x is either found in $A[n]$ or not found at all. The latter case gives us the *worst-case complexity* of the sequential search procedure. This tells us first that no matter what the values in the input data are, we are guaranteed never to need more than n look-ups, and second that this upper bound is in fact attained for certain possible input values.

In actual practice, the cost of sequentially searching for x in the array A will likely be a number of look-ups that falls somewhere between the two extremes of 1 and n. If we assume that x is equally likely to be stored in any of the n locations $A[1]$, $A[2]$, ..., $A[n]$, and that in addition, x is always

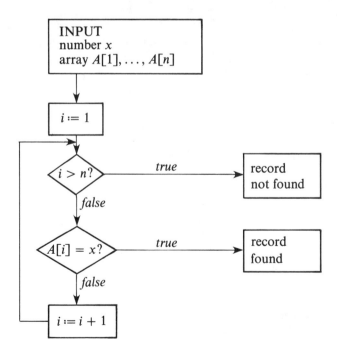

Figure 28 Flow diagram for sequential search of an array.

found somewhere in the array A, then on the average x will be $(1/n)$th of the time in each of the n locations. With these assumptions the average number of look-ups will be

$$\frac{1}{n}(1 + 2 + \cdots + n) = \frac{1}{n}\left(\sum_{1 \le i \le n} i\right) = \frac{1}{n} \cdot \frac{n(n + 1)}{2} = \frac{n + 1}{2}.$$

This gives us the *average complexity* of the sequential search procedure. It tells us that, under the stated assumptions, every search of the array A will on the average look up about half the entries.

The preceding solution can be improved considerably if the array is sorted, i.e., if $A[i] < A[j]$ for all $i < j$. In this case search for an entry, "dictionary" style: (1) Open the dictionary to the center; if the search key x is there, stop; (2) if x precedes the entries in the center page, then only the first half of the dictionary need be searched, and open to the center of that remaining half (i.e., go to (1)); (3) if x follows the entries in the center page, then only the second half of the dictionary need be searched, and open to the center of that remaining half. In this process, called *binary search* because it successively splits the dictionary in two — either the key x is located by a particular "probe," or else the remaining interval of entries to be searched is halved.

The binary search of (sorted) array A for an entry equal to x may be programmed as in Figure 29. In this algorithm we have used the operation **div** (integer division) introduced in Section 1.1. The effect of the instruction $k := (i + j)$ **div** 2 is to assign to k the value $\lfloor (i + j)/2 \rfloor$, the midpoint of the interval running from i to j.

How do we know that the above algorithm works correctly? In the case of sequential search the flow chart was somewhat simpler, and we left it to the reader to verify that it worked as claimed. Perhaps a close examination of the flow chart for binary search, given above, will also convince the reader of its correctness. But if we ask for a more formal argument that the program behaves as intended, then we are asking for a *proof of correctness* — a rigorous demonstration that the algorithm outputs the correct value for all possible inputs. The study of correctness proofs is a major topic in theoretical computer science, which will not however occupy us here. (A textbook account of the subject is given by S. Alagić and M. A. Arbib in their book, *The Design of Well-Structured and Correct Programs*, Springer-Verlag, 1978.) Nevertheless, all of the algorithms mentioned in this section can be rigorously proved to be correct.

Let us analyze the complexity of binary search. Specifically, we want to count the number of entries that must be examined in array A before an output is given. It will be useful to describe the sequence of comparisons that binary search goes through in the form of a binary tree. We use the notation "$x: A[k]$" to indicate a comparison between x and the contents of the kth array cell.

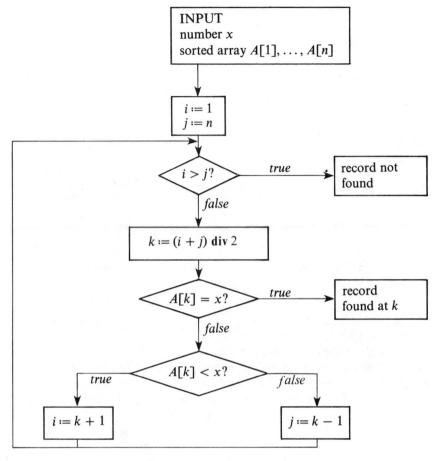

Figure 29   Flow diagram for binary search of a sorted array.

The first comparison is $x: A[\lfloor(n + 1)/2\rfloor]$, which we place at the root of the tree

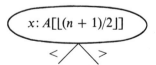

The left and right subtrees are the tree structures of the succeeding comparisons when $x < A[\lfloor(n + 1)/2\rfloor]$ and $x > A[\lfloor(n + 1)/2\rfloor]$ respectively. For example the tree corresponding to a binary search for x in a sorted array A containing ten entries is given in Figure 30.

Note that whenever there are no further comparisons that can be made in a subtree, we indicate this by a leaf marked with a square node. The resulting

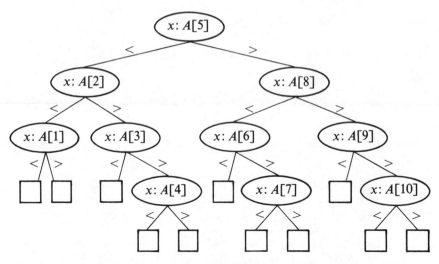

Figure 30   Decision tree for binary search where a number x is sought in a sorted array containing 10 elements.

binary tree is such that all internal nodes are ovals, and each has two children, and all leaves are square nodes.

The tree just defined is called a binary decision tree. Each path through this tree, from the root to one of the leaves, represents a sequence of comparisons in the binary search method. If x is present, then the algorithm will stop at one of the circular nodes. If x is not present, the algorithm will terminate at one of the square nodes.

Having placed square nodes in the fashion described above, a binary decision tree is always of height 1 or more. Moreover, if N is the number of internal nodes in such a tree, and if N_L and N_R are the number of nodes in the left and right subtrees, then $N_L = N_R$ or $N_L = N_R - 1$. This is so because the internal nodes in the subtrees cover the search interval less the center point, which is chosen as $\lfloor n + 1 \rfloor / 2$ for a length n interval, thus dividing the two subtrees as evenly as possible.

1 Lemma.

1. *The left and right subtrees, T_L and T_R, of any node in a binary decision tree are such that* $\text{height}(T_L) = \text{height}(T_R)$ *or* $\text{height}(T_L) = \text{height}(T_R) - 1$.
2. *The total number n of internal (i.e., oval) nodes in a binary decision tree of height $h \geq 1$ is such that $2^{h-1} \leq n < 2^h$.*
3. *In a binary decision tree of height $h \geq 1$, all square nodes are either at level $(h - 1)$ or level h, and all oval nodes are at levels $0, 1, 2, \ldots, (h - 1)$.*

PROOF. By induction on the height h, we prove simultaneously parts (1), (2), and (3). That is, we shall first prove the basis step for each part of the

CHAPTER 4

Switching Circuits, Proofs, and Logic

The two-element set plays a vital role in computer science, both as the bits 0 and 1 of binary arithmetic (corresponding to the two stable states of the basic subcomponents of computer circuitry), and as the two truth values, T (true) and F (false), of the logical analysis of circuitry, and the tests of programs. Section 4.1 introduces the basic operations of negation, conjunction, and disjunction on truth values, and shows how these operations can be used to express Boolean functions in normal form. We briefly discuss the use of components corresponding to these basic operations in building up computer circuits for addition. Section 4.2 looks at proof techniques from two perspectives. The first makes explicit the various "everyday" techniques we use in proving theorems in this book. The second gives a formal discussion of the notion of proof in propositional logic, and closes with a brief introduction to the quantifiers of predicate logic.

4.1 Truth Tables and Switching Circuits

A *proposition* is a statement that is either true or false. "My aunt lives in Boston" is a proposition; "The capital of Massachusetts" is not. If a proposition is true, we say its *truth value* is T (short for true); while if the statement is false, we say its truth value is F (short for false). The set $\mathscr{B} = \{T, F\}$, which we met in Section 1.1, is then the set of truth values.

Given a statement we can form its *negation*, or denial. "All men are mortal" has truth value T; its negation is "Not every man is mortal," the English form of "NOT (all men are mortal)" and has truth value F. We may thus represent negation by a map $\{T, F\} \rightarrow \{T, F\}$ which assigns to the truth value of a proposition (denoted by some letter, p), the truth value of the negation of that proposition (denoted $\bar{p}$ or $\neg p$ or $\text{NOT}(p)$). Clearly, $\bar{p}$ is defined by the *truth table*

p	$\bar{p}$
T	F
F	T

which pairs each truth value of p with the corresponding truth value of $\bar{p}$.

Another way of combining statements is *conjunction*. An example is the proposition "Harry is young AND Mary is old," which is true just in case both "Harry is young" and "Mary is old" are true. We are thus led to define a map

$$\wedge: \{T, F\}^2 \rightarrow \{T, F\}, \qquad (p, q) \mapsto p \wedge q$$

whose values are defined by the truth table

p	q	$p \wedge q$
T	T	T
T	F	F
F	T	F
F	F	F

Yet another way of combining two statements is to use the word "or." But there are two different ways of defining "or" by means of truth tables. The first case is the *exclusive or*; in this case (A or B) is true if either A or B is true, but not if both are. An example of the use of the exclusive "or" is the proposition "He took the train *or* he took the plane." The corresponding function $\oplus: \{T, F\}^2 \rightarrow \{T, F\}$ is given by the truth table

p	q	$p \oplus q$
T	T	F
T	F	T
F	T	T
F	F	F

The second case is the *inclusive or* — (A OR B) — which is true if either of A or B is true, or if both are true. An example is the proposition "The weather will be fine on Tuesday or Wednesday," considered as an abbreviation for "The weather will be fine on Tuesday OR the weather will be fine on Wednesday," which includes the possibility that the weather may be fine on both

Tuesday and Wednesday. The corresponding $\{T, F\}^2 \to \{T, F\}$, $(p, q) \mapsto$ $p \vee q$ is given by

p	q	$p \vee q$
T	T	T
T	F	T
F	T	T
F	F	F

The choice of " $\vee$ " comes from the fact that *vel* is the Latin word for the inclusive or; then the " $\wedge$ " is the " $\vee$ " upside down. This "upside down" relationship between the symbols for AND and OR is reflected logically by the equation

$$\overline{p \wedge q} = \bar{p} \vee \bar{q}. \tag{1}$$

The truth tables below show that this equivalence does indeed hold.

p	q	$\overline{p \wedge q}$		p	q	$\bar{p} \vee \bar{q}$
T	T	F		T	T	F
T	F	T		T	F	T
F	T	T		F	T	T
F	F	T		F	F	T

Another simple truth table manipulation shows us that

$$\bar{\bar{p}} = p. \tag{2}$$

Let us use these two facts to evaluate $\overline{p \vee q}$:

$$\overline{p \vee q} = \overline{\bar{\bar{p}} \vee \bar{\bar{q}}} \qquad \text{by two applications of (2)}$$
$$= \overline{\overline{\bar{p} \wedge \bar{q}}} \qquad \text{on applying (1) to } \bar{p} \text{ and } \bar{q}$$
$$= \bar{p} \wedge \bar{q} \qquad \text{by applying (2) again.}$$

We thus have *De Morgan's Laws*: For all pairs of truth values p and q,

$$\overline{p \wedge q} = \bar{p} \vee \bar{q}$$
$$\overline{p \vee q} = \bar{p} \wedge \bar{q}.$$

Thus $\wedge$ can be replaced by a composite of $\vee$ and $^-$; and $\vee$ can be replaced by a combination of $\wedge$ and $^-$. However, $^-$ *cannot* be replaced by a combination of $\wedge$ and $\vee$. This is because if we think of T as representing 1 and F as representing 0, then $1 \wedge 1 = 1$ and $1 \vee 1 = 1$, but $^-$ converts 1 to 0, something no combination of $\wedge$'s and $\vee$'s can do. (This is not a rigorous proof. See Exercise 4 for a sketch of how to proceed more formally.)

THE SWITCHING CIRCUIT VIEWPOINT

We now examine how each propositional function introduced above may be thought of as a switching function of the kind used in building computer circuitry. We replace T by 1 and F by 0, respectively, to get the "bits" of a computer's binary arithmetic. In what follows we may, for example, think of 0 as representing a low voltage on a line in the circuit, and of 1 as representing a high voltage.

A **NOT**-*gate* is an *inverter*:

p	$\bar{p}$
0	1
1	0

An **AND**-*gate* emits a 1 only if both input lines carry a 1:

$$0 \wedge 0 = 0 \quad 0 \wedge 1 = 0$$
$$1 \wedge 0 = 0 \quad 1 \wedge 1 = 1.$$

An **OR**-*gate* emits a 1 if at least one of the input lines carries a 1:

$$0 \vee 0 = 0 \quad 0 \vee 1 = 1$$
$$1 \vee 0 = 1 \quad 1 \vee 1 = 1.$$

If we think of inverters as "beads" which we can "slide" along a line until they rest against an AND-gate or OR-gate, we can represent De Morgan's laws by:

$$p \wedge q = \bar{p} \vee \bar{q}$$

$$p \vee q = \bar{p} \wedge \bar{q}$$

We shall explore this switching function interpretation in more detail later in this section. For the present let us consider propositional logic more closely from the numerical point of view. We start with the following simple relationships.

$$\bar{p} = (1 - p) \quad \text{since } 1 - 0 = 1, 1 - 1 = 0.$$

$$p \vee q = \max(p, q) \quad \text{the larger of } p \text{ and } q.$$

$$p \wedge q = \min(p, q) \quad \text{the smaller of } p \text{ and } q.$$

Let us also look at $p \oplus q$ and see what it means numerically.

p	q	$p \oplus q$
0	0	0
0	1	1
1	0	1
1	1	0

$p \oplus q$ is precisely *the sum of p and q modulo 2* — i.e., add p and q as ordinary numbers, and take the remainder after division by 2. Recall from Section 1.1 the notation n **mod** m for the "remainder when n is divided by m," also referred to as "n modulo m," which is the number r (the *remainder*) such that there exists some number q (the *dividend*) satisfying

$$n = q * m + r \quad \text{with } 0 \le r \le m - 1.$$

When the *modulus* m is 2, we have

$$(0 + 0) \bmod 2 = 0 \bmod 2 = 0 = 0 \oplus 0$$

$$(0 + 1) \bmod 2 = 1 \bmod 2 = 1 = 0 \oplus 1$$

$$(1 + 0) \bmod 2 = 1 \bmod 2 = 1 = 1 \oplus 0$$

$$(1 + 1) \bmod 2 = 2 \bmod 2 = 0 = 1 \oplus 1.$$

In general, given any natural number $m > 1$, we can define $\mathbf{Z}_m$ to be the set $\{0, 1, 2, \ldots, m - 1\}$ which we refer to as the set of integers modulo m. We have the map

$$\mathbf{Z} \to \mathbf{Z}_m, \qquad n \mapsto n \bmod m$$

which sends each integer n in $\mathbf{Z}$ to its (positive) remainder modulo m. This map can be visualized by thinking of $\mathbf{Z}$ as being wound into a helix, with m integers in every revolution, and then projecting the whole helix down onto the bottom ring — the single turn $\{0, 1, \ldots, m - 1\}$, as shown in Figure 31 for the case $m = 4$.

We can define a successor function $\sigma \colon \mathbf{Z}_m \to \mathbf{Z}_m$ by $\sigma(n) = (n + 1) \bmod m$, that is,

$$\sigma(n) = \begin{cases} 0 & \text{if } n = m - 1 \\ n + 1 & \text{if } 0 \le n < m - 1 \end{cases}$$

and we can define addition modulo m, $\oplus_m \colon \mathbf{Z}_m^2 \to \mathbf{Z}_m$ by

$$n \oplus_m n' = (n + n') \bmod m$$

where $+$ is the usual addition defined on the integers. Note, then, that we use $\oplus$ as an abbreviation for $\oplus_2$.

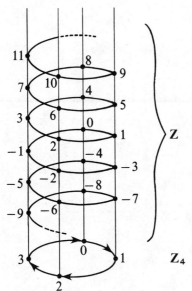

Figure 31   Mapping **Z** onto $\mathbf{Z}_4$.

Returning to logic gates, let us see how to build $\oplus$ using **AND, OR**, and **NOT** gates. Recall first that $p \oplus q$ is true just in case $p \vee q$ is true save with one exception, namely when $p \wedge q$ is true. That is $p \oplus q$ is true if $p \vee q$ is true and $p \wedge q$ is false:

$$p \oplus q = (p \vee q) \wedge \overline{(p \wedge q)}. \tag{3}$$

Let us check this by formal substitution in truth tables:

p	q	$p \vee q$	$\overline{p \wedge q}$	$(p \vee q) \wedge \overline{(p \wedge q)}$
0	0	0	1	0
0	1	1	1	1
1	0	1	1	1
1	1	1	0	0

We compute the third and fourth columns directly, and then use the $\wedge$-rules to find the fifth column, which is indeed the same as the column for $\oplus$, thus verifying (3). In other words, we have the following $\oplus$-circuit:

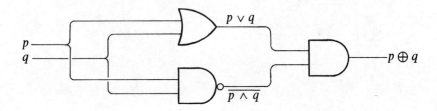

NORMAL FORMS

A common technique in mathematics is to reduce complicated expressions to some *normal form*.

Perhaps the most familiar example of a normal form is the expression of a fraction in its lowest terms

$$\frac{1}{2} \text{ is the normal form for } \frac{1}{2}, \frac{2}{4}, \frac{3}{6}, \ldots, \frac{117}{334}, \text{ etc.}$$

$$\frac{7}{9} \text{ is the normal form for } \frac{7}{9}, \frac{14}{18}, \frac{21}{27}, \ldots, \frac{136332}{175284}, \text{ etc.}$$

Another normal form — so-called scientific notation — allows us to express any nonzero decimal number in the form $a \times 10^b$ where a has only one digit to the left of the decimal point, and b is an integer, so that

$$1.375 \times 10^6 \text{ is the normal form for } 1375000$$

$$7.215 \times 10^{-4} \text{ is the normal form for } .0007215.$$

We now examine normal forms for formulas built up using $\wedge$, $\vee$, $\bar{}$, and propositional letters. Consider the formula

$$(p \vee \bar{q}) \wedge \overline{(r \vee q)}. \tag{4}$$

This expression has the following truth table

p	q	r	$(p \vee \bar{q}) \wedge \overline{(r \vee q)}$
1	1	1	0
1	1	0	0
1	0	1	0
1	0	0	1
0	1	1	0
0	1	0	0
0	0	1	0
0	0	0	1

The *disjunctive normal form* for this formula is

$$(p \wedge \bar{q} \wedge \bar{r}) \vee (\bar{p} \wedge \bar{q} \wedge \bar{r}). \tag{5}$$

That is, formula (4) and formula (5) are equivalent formulas since they have the same truth tables, and (5) is in *disjunctive normal form*: it is a disjunction of conjuncts of variables and negations of variables, and each conjunct mentions every variable exactly once.

Expression (5) was obtained in the following way. Examining the truth table above, we see that exactly two of its rows yield a true value for the original formula. The first row occurs when p is true and q and r are false — a set of conditions we can represent by the expression $(p \wedge \bar{q} \wedge \bar{r})$. This expression is false whenever these conditions are not exactly met. Similarly, $(\bar{p} \wedge \bar{q} \wedge \bar{r})$ represents the second true row of the truth table. Since *either* pattern of values to the variables is sufficient to yield a true row, the disjunction of these two conditions, formula (5) above, is the desired expression.

Our result shows that any formula built up from $\wedge$, $\vee$, and $^{-}$ has an equivalent disjunctive normal form. But in fact a more general result, yielding a disjunctive normal form for any "truth-table" or "Boolean function" is also true.

1 Definition. A *Boolean function of n arguments* is a map

$$f: \{0, 1\}^n \to \{0, 1\}$$

which assigns a value $f(p_1, \ldots, p_n)$ of 0 or 1 to each n-tuple $(p_1, \ldots, p_n)$ of 0's and 1's.

There are 2^n elements in $\{0, 1\}^n$. We can therefore write a truth table for f with 2^n lines, one for each n-tuple in $\{0, 1\}^n$, the n-tuple $(p_1, \ldots, p_n)$ being followed by the value $f(p_1, \ldots, p_n)$ that f assigns to that particular n-tuple. Hence, there are 2^{2^n} different n-ary truth tables. (Alternatively, recall from **1.3.9** that there are $|B|^{|A|}$ distinct maps from A to B. Here $A = \{0, 1\}^n$ with $|A| = 2^n$ and $B = \{0, 1\}$ with $|B| = 2$.) In the course of showing that every Boolean function of n arguments has a disjunctive normal form, we shall also show that such a function can be built up just using $\wedge$, $\vee$, $^{-}$, and propositional letters $p_1, \ldots, p_n$.

Consider first the operation of conjunction. It is clear that it is associative; that is,

$$p \wedge (q \wedge r) = (p \wedge q) \wedge r$$

since both sides take the value 1 iff *all* of p, q, and r take the value 1. We can thus remove the parentheses and write expressions like

$$q_1 \wedge q_2 \wedge \cdots \wedge q_n \tag{6}$$

as short for any of the equivalent repeated conjunctions such as

$$((\cdots(q_1 \wedge q_2) \wedge \cdots) \wedge q_n).$$

We also abbreviate (6) to $\bigwedge_{1 \le i \le n} q_i$ or $\bigwedge_{i \in I} q_i$ for $I = \{1, 2, \ldots, n\}$. In the latter case I is called the index set for the conjunction. Similarly, given an index set I with one proposition q_i for each $i \in I$, we may use the notation

$$\bigvee_{i \in I} q_i$$

for the repeated disjunction which takes the value 1 iff *at least one* of the q_i's is 1. Now, consider the notation

$$p^0 \text{ as short for } \bar{p}$$

$$p^1 \text{ as short for } p.$$

Then note that $0^0 = 1$ and $0^1 = 0$, while $1^0 = 0$ and $1^1 = 1$. We thus see that $p^x = 1$ iff $p = x$, for $p, x \in \{0, 1\}$. Let us fix, then, an n-tuple of Boolean values, $(x_1, \ldots, x_n) \in \{0, 1\}^n$ and use it to define a Boolean function:

$$(p_1, \ldots, p_n) \to p_1^{x_1} \wedge p_2^{x_2} \wedge \cdots \wedge p_n^{x_n}.$$

We see that this function takes the value 1 just in case $p_i^{x_i} = 1$ for every $1 \le i \le n$; i.e., just in case $p_i = x_i$ for every $1 \le i \le n$. The truth table of the Boolean function just defined has all the entries of its right-hand side column equal to 0, except for a single entry in the row corresponding to the fixed n-tuple $(x_1, \ldots, x_n)$.

We can see this directly by inspecting the truth tables for $n = 2$, one for every function of the form $(p, q) \mapsto p^x \wedge q^y$:

p	q	$\bar{p} \wedge \bar{q}$
0	0	1
0	1	0
1	0	0
1	1	0

$(x, y) = (0, 0)$

p	q	$\bar{p} \wedge q$
0	0	0
0	1	1
1	0	0
1	1	0

$(x, y) = (0, 1)$

p	q	$p \wedge \bar{q}$
0	0	0
0	1	0
1	0	1
1	1	0

$(x, y) = (1, 0)$

p	q	$p \wedge q$
0	0	0
0	1	0
1	0	0
1	1	1

$(x, y) = (1, 1)$.

Any other Boolean function of two arguments which has in the right-hand side column of its truth table more than one entry equal to 1 can now be obtained as a combination of the four defined above. Consider for instance the truth table for $p \oplus q$: it has a 1 in the $(0, 1)$ and $(1, 0)$ lines; i.e., it has value 1 just in case $p^0 \wedge q^1 = \bar{p} \wedge q$ is true or $p^1 \wedge q^0 = p \wedge \bar{q}$ is true, and so we have

$$p \oplus q = (\bar{p} \wedge q) \vee (p \wedge \bar{q})$$

as an alternative way of expressing $\oplus$ in terms of $\wedge$, $\vee$ and $\bar{\ }$.

2 Example. Let $f: \{0, 1\}^3 \to \{0, 1\}$ be such that $f(p, q, r) = 0$ unless (p, q, r) has one of the values $(1, 0, 1)$ or $(0, 0, 1)$ or $(0, 1, 0)$. Then

$f(p, q, r) = (p \wedge \bar{q} \wedge r) \vee (\bar{p} \wedge \bar{q} \wedge r) \vee (\bar{p} \wedge q \wedge \bar{r})$, which yields the switching function:

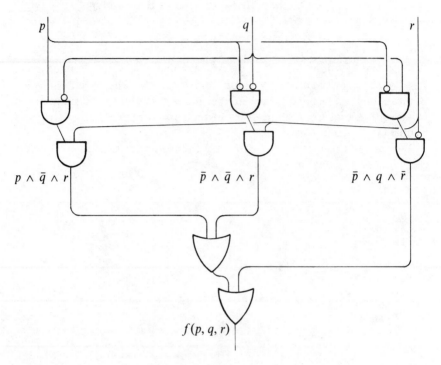

In the same way we can build up *any* Boolean function:

3 Theorem. *Given any Boolean function* $f: \{0, 1\}^n \to \{0, 1\}$, *we may express* $f(p_1, \ldots, p_n)$ *as the disjunction of those conjuncts* $p_1^{x_1} \wedge \cdots \wedge p_n^{x_n}$ *for which* $f(x_1, \ldots, x_n)$ *equals* 1:

$$f(p_1, \ldots, p_n) = \bigvee_{\{(x_1, \ldots, x_n) | f(x_1, \ldots, x_n) = 1\}} p_1^{x_1} \wedge \cdots \wedge p_n^{x_n}$$

$$= \bigvee_{\{(x_1, \ldots, x_n) | (x_1, \ldots, x_n) \in \{0, 1\}^n\}} f(x_1, \ldots, x_n) \wedge p_1^{x_1} \wedge \cdots \wedge p_n^{x_n}.$$

PROOF. This is a generalization of the preceding discussion. When we choose specific values for $(p_1, \ldots, p_n)$, the only term of the disjunction which is non-zero satisfies $p_i = x_i$ for all $1 \leq i \leq n$, and so the value of the disjunction is 1 if and only if $f(p_1, \ldots, p_n) = 1$. $\qquad\square$

The **disjunctive normal form theorem**, Theorem **3** above, shows that any Boolean function can be built up using only the operations $\wedge$, $\vee$, and $^-$.

Note that each conjunct in the disjunctive normal form for a Boolean function of n arguments mentions every one of the variables (i.e., propositional letters) in the set $\{p_1, \ldots, p_n\}$. Such a conjunct is called a *minterm*. The characteristic property of a minterm is that it assumes the value 1 for exactly one assignment of values to the n variables. We can easily see that a minterm is unique (up to a reordering of the n variables), in that no other minterm assumes the value 1 for the same combination of the n arguments. This immediately leads to the following fact.

4 Corollary. *The disjunctive normal form of a Boolean function $f(p_1, \ldots, p_n)$ is unique, up to a reordering of the n variables in each conjunct and up to a reordering of the conjunct in the disjunctive normal form.*

PROOF. We already pointed out that the value of a conjunct is not affected by the order in which its variables are mentioned, since the operation $\wedge$ is commutative, i.e., $p \wedge q = q \wedge p$. Likewise, the value of a disjunctive normal form is not affected by the order of its disjuncts, since the operation $\vee$ is commutative. □

Another normal form for a Boolean function $f: \{0, 1\}^n \to \{0, 1\}$ is the so-called *conjunctive normal form*, which uniquely expresses $f(p_1, \ldots, p_n)$ as a conjunction of disjuncts. The existence and uniqueness of the conjunctive normal form can be easily established in a manner similar to our discussion of the disjunctive normal form above (see Exercise 2). However, once we have found the disjunctive normal form for f, we may also use De Morgan's laws to determine the conjunctive normal form of f, as shown in the next example.

5 Example. This example continues Example 2. The negation of f is the function $\bar{f}(p, q, r)$ which has value 1 at the following values of its arguments: $(0, 0, 0), (0, 1, 1), (1, 0, 0), (1, 1, 0)$, and $(1, 1, 1)$. Hence, the disjunctive normal form of $\bar{f}$ is:

$$\bar{f}(p, q, r) = (\bar{p} \wedge \bar{q} \wedge \bar{r}) \vee (\bar{p} \wedge q \wedge r) \vee (p \wedge \bar{q} \wedge \bar{r}) \vee (p \wedge q \wedge \bar{r})$$
$$\vee (p \wedge q \wedge r)$$

which includes all the minterms not included in the disjunctive normal form of f. Negating $\bar{f}$ and applying De Morgan's laws, we now get

$$f(p, q, r) = \bar{\bar{f}}(p, q, r) \,.$$
$$= \overline{(\bar{p} \wedge \bar{q} \wedge \bar{r})} \wedge \overline{(\bar{p} \wedge q \wedge r)} \wedge \overline{(p \wedge \bar{q} \wedge \bar{r})} \wedge \overline{(p \wedge q \wedge \bar{r})}$$
$$\wedge \overline{(p \wedge q \wedge r)}$$
$$= (p \vee q \vee r) \wedge (p \vee \bar{q} \vee \bar{r}) \wedge (\bar{p} \vee q \vee r) \wedge (\bar{p} \vee \bar{q} \vee r)$$
$$\wedge (\bar{p} \vee \bar{q} \vee \bar{r})$$

which is the conjunctive normal form for f. Each of the disjuncts in the conjunctive normal form of f is called a *maxterm*, because it mentions each of the variables p, q, and r, and assumes the value 1 for all but one of the assignments of values to the three variables.

Generally, then, if the disjunctive normal form of the function $f: \{0, 1\}^n \rightarrow \{0, 1\}$ is known, then the disjunctive normal form of the negation of f, $\bar{f}: \{0, 1\}^n \rightarrow \{0, 1\}$, will consist of the disjunction of the remaining minterms which do not appear in the disjunctive normal form of f. Now, since $f = \bar{\bar{f}}$, we can obtain the conjunctive normal form of f by repeated applications of De Morgan's laws to the disjunctive normal form of $\bar{f}$.

FUNCTIONALLY COMPLETE OPERATIONS

A set of operations is said to be *functionally complete* if every Boolean function can be expressed entirely by means of operations from this set. Theorems **3** and **4** above show that the set $\{\wedge, \vee, \bar{\ }\}$ is functionally complete.

We have seen that De Morgan's laws let us either build $\wedge$ from $\vee$ and $\bar{\ }$, or build $\vee$ from $\wedge$ and $\bar{\ }$. Hence, again using Theorem **3** or **4**, we have

6 Fact. *The sets* $\{\vee, \bar{\ }\}$ *and* $\{\wedge, \bar{\ }\}$ *are functionally complete.*   □

In fact, there is a Boolean function from which all other Boolean functions can be composed. It is the Sheffer stroke $p|q$:

p	q	$p|q$
0	0	1
0	1	1
1	0	1
1	1	0

We see that $p|q = \overline{p \wedge q}$, and so the Sheffer stroke is often called **NAND**, short for **NOT-AND**. It can be realized by a **NAND**-*gate*

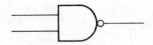

Given the Sheffer stroke, we recapture $^-$, $\wedge$ and $\vee$ as follows:

$$\bar{p} = \overline{p \wedge p} = p|p$$
$$p \wedge q = \overline{\overline{p \wedge q}} = \overline{p|q} = (p|q)|(p|q)$$
$$p \vee q = \overline{\bar{p} \wedge \bar{q}} = \overline{(p|p) \wedge (q|q)} = (p|p)|(q|q).$$

We thus have the following

7 Fact. *The Sheffer stroke is functionally complete.*   □

There is still another operation, a **NOR** operation, $(p \downarrow q) = \overline{(p \vee q)}$, the "dagger" $\downarrow$ in terms of which all other Boolean functions can be defined, which we shall leave to the reader's consideration (Exercise 1).

ADDING BITS

As an example of the logical design of a circuit which can be used as part of a computer, consider a *half-adder* (we shall meet the full-adder below), which takes two bits, a and b, as input and returns a carry bit c and sum bit s as output. The pair cs in that order is the binary sum of a and b.

a	b	c	s
0	0	0	0
0	1	0	1
1	0	0	1
1	1	1	0

We recognize that $c = a \wedge b$ while, using the disjunctive normal form, $s = a \oplus b = (\bar{a} \wedge b) \vee (a \wedge \bar{b})$. Thus we can build our half-adder using a total of four **AND**- and **OR**-gates as follows:

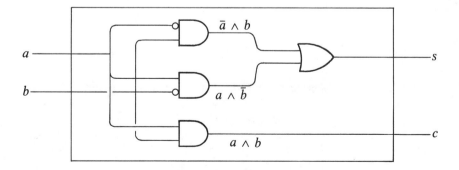

However, if we recall the formula $a \oplus b = (a \vee b) \wedge (\overline{a \wedge b})$, we can build the half-adder with one less gate.

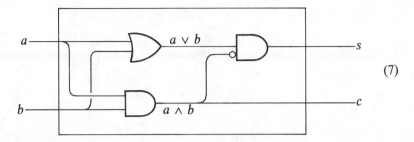

(7)

These two circuits exemplify the general observation that while disjunctive normal form tells us how to build any switching functions using **AND-**, **OR-**, and **NOT**-gates, it does *not* in general give us the *minimal* circuit, i.e., the one with the smallest number of gates. The problem of finding minimal circuits was an active area of research in the 50's and 60's. But computer circuitry is no longer built by wiring individual components together. Instead, integrated circuits containing the equivalent of tens of thousands of gates can be mass produced by a process akin to printing photographs. (See, for example, the article on "Microcomputers" by A. G. Vacroux in the May 1975 *Scientific American*, which gives a useful account of how the entire circuitry of a computer can be made from just a few specially designed "chips.") As a result, minimization problems have been replaced by layout problems — how do we lay out components to reduce the length of con-nections of the number of crossovers? — and these are usually handled by computer-assisted design techniques outside the scope of this book.

Returning now to the binary addition problem: We want to design a *full-adder* which can take the bits a and b together with a carry bit c' to return the sum bit s and carry bit c.

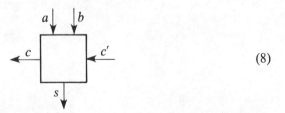

(8)

The desired sum bit s is clearly $(a \oplus b) \oplus c'$. But what of the carry bit c as a function of a, b, and the "input carry" c'? We want c to be 1 just in case *at least* two of a, b, and c' are 1. But $(a \oplus b) \wedge c'$ is 1 if only a and c' are 1 or only b and c' are 1; while $a \wedge b$ is true if only a and b are 1, or if all a, b, and c' are 1. Thus

$$c = [(a \oplus b) \wedge c'] \vee (a \wedge b).$$

Thus we can realize the full-adder of (8) (changing the position of input and output lines) by combining two half-adders with an **OR**-gate as follows:

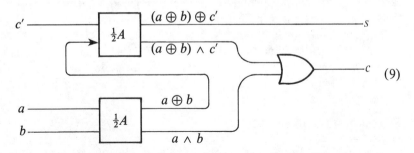

(9)

Given a collection of full-adders, we can add binary numbers of any length, as in the circuit below for adding 4-bit numbers $(b_4 b_3 b_2 b_1)$ and $(a_4 a_3 a_2 a_1)$ to yield a 5-bit number $(c_4 s_4 s_3 s_2 s_1)$:

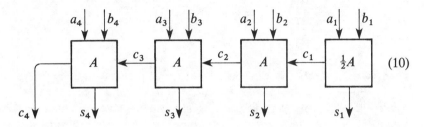

(10)

A problem of great interest to computer scientists is to determine the complexity of *switching functions* — to find lower bounds on the speed with which a function can be computed, and to seek circuits which can compute the function in close to minimal time. As an example of this, consider the time required to add two binary numbers with circuit (10). Take as our time unit the time required for an **AND**-gate or **OR**-gate (with or without inverters) to achieve a stable output level after the input values have been changed. Then circuit (7) for a half-adder will take 2 time units to stabilize its output after a change in input values; and the full-adder of (8) will take 5 time units to stabilize. Consider, then, an n-bit adder such as that in (10) for $n = 4$. Circuit (10) is inherently *serial*, in that the $(j + 1)$th full-adder cannot begin to compute the correct outputs until it receives a stable carry bit c_j from the preceding full-adder. Thus the total time to achieve a stable output from such an n-bit adder is

$$2 + 5(n - 1) \simeq 5n$$

where the notation $\simeq$ indicates that the two sides are almost equal, and that the difference becomes less and less important as n gets larger. It may appear to the reader that no n-bit adder could be designed without requiring a time proportional to the number n of bits, for the propagation of carry bits seems to be the limiting factor. However, there is an ingenious design called the *carry look-ahead adder* which uses *parallelism* based on "guessing carry bits and patching up later" which has a time requirement proportional to the logarithm of n. For 64 bits, the look-ahead adder is about 9 times faster, yet only requires about 50% more gates than the conventional adder like (10). (For actual circuit details, see Sections 11.4 to 11.6 of F. J. Hill and G. R. Peterson: "Digital Systems: Hardware Organization and Design," Wiley, 1973.)

THRESHOLD LOGIC UNITS

In 1943, Warren McCulloch and Walter Pitts offered a formal model of the neuron (the basic cellular component of the brain) as a switching function $\{0, 1\}^n \to \{0, 1\}$ defined in the following way:

$$(11)$$

There is a real number θ called the *threshold*, and for each input line there is a *weight* w — we say the input is *excitatory* if $w > 0$ and *inhibitory* if $w < 0$. Given a binary input vector $(p_1, \ldots, p_n)$ we form the real number sum $\sum_{i=1}^{n} w_i p_i$, and decree that the output is 1 just in case this sum reaches threshold:

$$p = \begin{cases} 1 & \text{if } \sum_{i=1}^{n} w_i p_i \geq \theta \\ \\ 0 & \text{if } \sum_{i=1}^{n} w_i p_i < \theta. \end{cases} \tag{12}$$

Such a component is referred to as a *McCulloch-Pitts Neuron* or a *Threshold Logic Unit*. (Such units have been of great interest in the field of *pattern recognition* where researchers study how the weights w_i might be adjusted to cause a threshold logic unit to respond only to those inputs belonging to some "pattern." For a review of this field, see N. J. Nilsson: "Learning Machines," McGraw-Hill, 1965.)

We now see how to choose weights and threshold to build a **NOT**-gate, and an **AND**-gate. Against each line of the truth table, we place the inequality which the weights and threshold must satisfy according to (12).

p	$\bar{p}$	
0	1	yields $0 \geq \theta$
1	0	yields $w < \theta$

and we can satisfy these inequalities with $\theta = 0$, $w = -1$.

p	q	$p \wedge q$	
0	0	0	yields $0 < \theta$
0	1	0	yields $w_2 < \theta$
1	0	0	yields $w_1 < \theta$
1	1	1	yields $w_1 + w_2 \geq \theta$

(figure: threshold logic unit with inputs $p \to w_1$, $q \to w_2$, threshold θ, output $p \wedge q$)

and we can satisfy these inequalities with $\theta = 2$, $w_1 = w_2 = 1$.

However, we shall see that no single threshold logic unit can compute $p \oplus q$. For suppose we have

p	q	$p \oplus q$	
0	0	0	yielding $0 < \theta$
0	1	1	yielding $w_2 \geq \theta$
1	0	1	yielding $w_1 \geq \theta$
1	1	0	yielding $w_1 + w_2 < \theta$.

We could then deduce that

$$w_1 + w_2 \geq \theta + \theta > \theta \quad \text{since } \theta > 0$$

but this would contradict $w_1 + w_2 < \theta$. Hence no such choice of w_1, w_2 and θ is possible.

This is an example of the very important notion of an *impossibility proof* — the proof that a problem cannot be solved using a particular methodology.

EXERCISES FOR SECTION 4.1

1. Let $(p \downarrow q)$ denote the Boolean function realized by the **NOR**-*gate*

(a) Write the truth table for $(p \downarrow q)$
(b) Show how to express $\wedge$, $\vee$ and $^-$ by repeated application of $\downarrow$.

2. Given any Boolean function $f: \{0, 1\}^n \to \{0, 1\}$, prove directly (i.e., without applying De Morgan's laws to the disjunctive normal form of f) that f can be written in the following *conjunctive normal form*

$$f(p_1, \ldots, p_n) = \bigwedge_{\{(x_1, \ldots, x_n) | f(\bar{x}_1, \ldots, \bar{x}_n) = 0\}} p_1^{x_1} \vee p_2^{x_2} \vee \cdots \vee p_n^{x_n}.$$

3. Use **AND, OR,** and **NOT**-gates to build a circuit which realizes the function $f: \{0, 1\}^3 \rightarrow \{0, 1\}^2$ defined by the table:

p	q	r	$f_1(p, q, r)$	$f_2(p, q, r)$
0	0	0	0	1
0	0	1	1	0
0	1	0	0	1
0	1	1	1	1
1	0	0	0	0
1	0	1	1	0
1	1	0	1	0
1	1	1	0	1

4. Prove by induction on the number of gates that no circuit built entirely from **AND**-gates and **OR**-gates can realize $\bar{p}$, where the circuit has one input line carrying p, and the output line of one gate in the circuit should (but you will prove it cannot) bear the value $\bar{p}$.

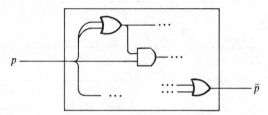

5. Choose weights w_1 and w_2 and threshold θ for a threshold logic unit to realize each of the following
   (a) $(p \downarrow q)$
   (b) $p \vee q$
   (c) $p \Leftrightarrow q$ defined by

p	q	$p \Leftrightarrow q$
0	0	1
0	1	0
1	0	0
1	1	1

or prove that no such realization exists.

6. Let **if** p **then** q **else** r be the Boolean function which returns the truth value of q if p is true, but returns the truth value of r if p is false. Verify that
   (a) **if** p **then** $\bar{p}$ **else** p equals *false*
   (b) **if** p **then** q **else** *true* equals $p \supset q$ which is defined by the truth table

p	q	$p \supset q$
T	T	T
T	F	F
F	T	T
F	F	T

(c) **if** p **then** q **else** *false* equals $p \wedge q$.

(d) **if** p **then** *false* **else** *true* equals $\bar{p}$.

What expression of this kind (called a 3-conditional) yields $p \vee q$?

(e) Show that the ternary connective **if-then-else** is complete.

(f) Give a Sheffer stroke representation of **if** p **then** q **else** r.

7. A more general form of conditional than that of Exercise 6 is given by

$$(\text{COND}(p_1 \, e_1)(p_2 \, e_2) \cdots (p_n \, e_n))$$

which is evaluated as follows: Find the first j such that p_j is true and return the value of e_j; if no such j exists, the result is undefined (a *partial* Boolean function!).

(a) Verify that **if** p **then** q **else** $r = (\text{COND}(p \, q)(T \, r))$

(b) Write down the disjunctive normal form for

(i) $(\text{COND}(p \wedge q \, q)(\bar{p} \, p)(\bar{q} \, q))$

(ii) $(\text{COND}(p \vee q \vee r \, \bar{q} \wedge \bar{r})(T \, r))$.

8. Notice that

$$\textbf{if } (\textbf{if } p \textbf{ then } q \textbf{ else } r) \textbf{ then } s \textbf{ else } n$$

is equivalent to

$$\textbf{if } p \textbf{ then } (\textbf{if } q \textbf{ then } s \textbf{ else } n) \textbf{ else } (\textbf{if } r \textbf{ then } s \textbf{ else } n).$$

Use this fact to devise an algorithm that transforms any **if-then-else** expression into an equivalent **if-then-else** expression in which the "**if**" expression is a simple variable.

9. By writing out truth tables show

(i) $(p \vee q) \vee s = p \vee (q \vee s)$

(ii) $(p \wedge q) \wedge s = p \wedge (q \wedge s)$

(iii) $(p \oplus q) \oplus s = p \oplus (q \oplus s)$

(iv) $[(p \wedge \bar{q}) \vee \bar{p}] \wedge \overline{[p \vee q]} = \bar{p} \wedge \bar{q}$.

10. How many Boolean functions $\{0, 1\}^n \to \{0, 1\}$ are there for $n = 1, 2,$ and 3? Display all the Boolean functions for $n = 2$ in a table as shown below. Label the functions encountered in the text with the appropriate symbol.

p	q	f_0	f_1	$\cdots$
0	0	0	$\cdots$	
0	1	0		
1	0	0		
1	1	0		

11. Give an example of a formula which is a disjunct of conjuncts and which is distinct from its disjunctive normal form.

12. Determine the function $f(p_1, p_2, p_3, p_4)$ realized by the following network

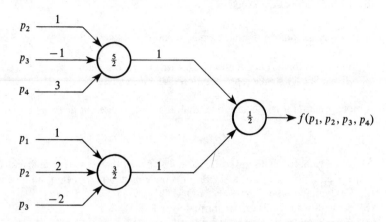

13. Determine which of the following Boolean functions are realized by threshold logic units
   (a) $f(p, q, r) = \bar{p}\,\bar{q}\,\bar{r} \vee \bar{p}\,\bar{q}\,r \vee \bar{p}\,q\,r$
   (b) $f(p, q, r) = p\bar{q} \vee q\bar{r}$
   (c) $f(p, q, r) = \bar{p}\,\bar{q}\,\bar{r} \vee \bar{p}\,q\,r \vee p\,\bar{q} \vee p\,q\,\bar{r}$
   (d) $f(p, q, r) = \bar{p}\,q \vee p\,q\,\bar{r}.$

14. Consider the following threshold network

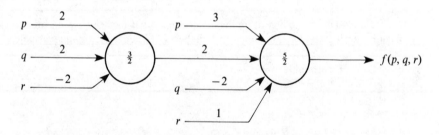

   (a) What is the truth table of the function $f(p, q, r)$?
   (b) Realize $f(p, q, r)$ with a single threshold logic unit.

15. A formula is in *biconditional form* if no connective other than $\Leftrightarrow$ is present. $p \Leftrightarrow q$; $p \Leftrightarrow (p \Leftrightarrow q)$, and q are in biconditional form. Given an example of a formula that is *not* equivalent to any formula in biconditional form. The truth table for $\Leftrightarrow$ is given below

p	q	$p \Leftrightarrow q$
T	T	T
T	F	F
F	T	F
F	F	T

4.2 Proving Theorems

In this section, we take a self-conscious look at the way in which we establish those "true statements" called theorems. In the first subsection we look at the proof techniques of everyday mathematics. We then take a quick look at quantifiers, which we use to formalize "for some" and "for all" statements. We close the section with a brief study of the formal notion of proof in logic.

PROOF TECHNIQUES IN "EVERYDAY MATHEMATICS"

Now that the reader has seen a number of proofs in the preceding sections, it is time for a more explicit look at some of the most useful proof techniques. (Note, too, the discussion of existence and constructive proofs on page 78.)

Proof by Induction. In Section 2.1 we presented the concept of proof by induction. Recall that to prove a property $P(n)$ of the natural numbers, we proceed in two steps:

1. *The Basis Step*: We demonstrate that $P(n)$ is true for $n = 0$. (If we wish to prove that $P(n)$ holds for all $n \geq m$, we begin with $n = m$.)
2. *The Induction Step*: We assume that $P(n)$ is true for $n = k$, and under this assumption demonstrate that $P(n)$ must be true for $n = k + 1$.

We also saw more general proofs by induction for properties of sets of strings and for sets of lists, and such proofs are also available for other sets of structured objects.

Proof by Exhaustion of Cases. A proof may be performed by demonstrating the validity of a statement for all possible cases. For example, in proving a property for all integers we might, as exemplified in the proof of Theorem 2 below, give three proofs showing that the property holds for all integers of the form $3m$, the form $3m + 1$, and the form $3m + 2$. Since all integers are of one of these three forms, this exhausts all possible cases.

Proof by Contradiction. The next proof technique we will consider is that of *proof by contradiction*. It is also called *reductio ad absurdum*: reduction to an absurdity. In this method, we use the property that a given statement is either true or false, but not both. We assume that the given statement is false, and then use definitions and facts (whether obvious or previously proven) to develop some statement which contradicts that which we know to be factual. Since the contradiction was developed by assuming that the given statement was false, and since we have used only known truths, our original assumption must have been incorrect. This then proves that the original statement is true.

At first sight, proof by contradiction may not appear to be significant, but is often the easiest and best proof technique. We gave a classical *reductio ad absurdum* in Section 1.1 when we presented Euclid's proof that there are infinitely many primes in **N**.

When a statement A implies a statement B, we say that B is a *necessary condition* for A. This is simply because it is "necessary" for B to be true when A is true. There is more than one way to say this:

1. A is true implies B is true.
2. If A is true, then B is true.
3. A can be true only if B is true.
4. A necessary condition for A to be true is that B is true.
5. It is necessary that B be true in order for A to be true.

The above five statements are equivalent, and can all be symbolized mathematically by $A \Rightarrow B$. To prove necessity, we simply start with A (the hypothesis), and present a series of supported statements until we conclude B (possibly by employing methods of contradiction).

Note that if $A \Rightarrow B$, it is *not* necessarily true that $B \Rightarrow A$ (which is the *converse* of $A \Rightarrow B$.) For example, suppose we take **Z** to be the set $\{\ldots, -3, -2, -1, 0, 1, 2, 3, \ldots\}$ of all integers, be they negative, zero, or positive; and then consider the two statements:

$$A_n \text{ which is true if } n \in \mathbf{Z} \text{ and } 0 \leq n \leq 2; \text{ and}$$
$$B_n \text{ which is true if } n \in \mathbf{Z} \text{ and } n^2 \leq 4.$$

Then $A_n \Rightarrow B_n$, but not $B_n \Rightarrow A_n$. Taking $n = -1$, we see a case in which B_n holds but A_n does not hold, thus showing that A_n cannot be a necessary condition for B_n.

If we denote the negation of A by $\bar{A}$ (so that A is false means $\bar{A}$ is true; and A is true means $\bar{A}$ is false), then it is useful to note that: proving $A \Rightarrow B$ is equivalent to proving the *contrapositive* $\bar{B} \Rightarrow \bar{A}$. To say that whenever A is true, then B must be true, is just to deny that B can be false when A is' true; in other words, it is easy to say that whenever B is false then A is false; i.e., that $\bar{B} \Rightarrow \bar{A}$. For example, the statement "If I live in Montreal, then I live in Canada" has as its contrapositive "If I don't live in Canada, I don't live in Montreal." We shall apply this technique in the second half of our proof of Theorem 1 below.

We previously mentioned that for the case $A \Rightarrow B$, we refer to B as a necessary condition for A. In a similar vein, we say that A is a *sufficient condition* for B. This is due to the fact that the truth of A is "sufficient" for us to conclude the truth of B. Since the statement of a necessary condition can take many forms, it's quite clear that the same is true for the statement of a sufficient condition. In fact, the three forms (1) to (3) given above constitute valid ways of saying that A is a sufficient condition for B. To be explicit, we would re-write statements (4) and (5) to the equivalent forms:

4′. *A sufficient condition for B to be true is that A is true.*
5′. *For B to be true, it is sufficient that A is true.*

NECESSARY AND SUFFICIENT CONDITIONS

We now discuss the most interesting case: That for which not only $A \Rightarrow B$, but in which we also have the converse, $B \Rightarrow A$ (B implies A) or $A \Leftarrow B$ (A is implied by B). We symbolize this situation by $A \Leftrightarrow B$. In words, we may say:

1. A is true if and only if B is true.
2. A necessary and sufficient condition that A be true is that B be true (or vice versa).

If one statement is both a necessary and sufficient condition for another, then the statements are equivalent even though they may be worded in completely different ways. Often it is more convenient to work with a necessary and sufficient condition of some property rather than directly with the original definition. Under this circumstance, we may wish to think of the condition as an alternative definition. Even in many definitions themselves, authors (including ourselves) will sometimes use the term "if and only if," although technically it is somewhat redundant.

Throughout the book we will either spell out "if and only if," shorten it to "iff," or use the symbol $\Leftrightarrow$ when dealing with necessary and sufficient conditions.

Finally, to prove $A \Leftrightarrow B$ (i.e., a necessary and sufficient condition), we must actually complete two proofs (the order is irrelevant): We must prove both $A \Rightarrow B$ and its converse $B \Rightarrow A$.

For some proofs, once it is demonstrated that $A \Rightarrow B$ it is just a matter of working backwards to show that $B \Rightarrow A$. Unfortunately, for a substantial number of proofs, it is impossible to proceed in this manner, and a completely different approach is required.

We close this section by proving a simple "if and only if" theorem, whose proof demonstrates most of the points we have made above.

1 Theorem. *A natural number is a multiple of 3 iff the sum of the digits in its decimal representation is a multiple of 3.*

It turns out that the proof of this theorem will be easier if we first prove a lemma. A *lemma* is a result we prove not for its own interest but because it *helps* us prove some theorems. (The German word for lemma is *Hilfsatz* — *Hilf* = help and *Satz* = theorem.)

2 Lemma. *Let $\langle n \rangle$ denote the decimal representation of the number $n \in \mathbf{N}$, and let $r(n)$ equal the sum of the digits in $\langle n \rangle$. Then $r(n + 3)$ differs from $r(n)$ by a multiple of 3.*

PROOF. We prove the result by exhausting two cases:

(I)   The last digit d of $\langle n \rangle$ is 0, 1, 2, 3, 4, 5 or 6. In this case we form $\langle n + 3 \rangle$ by changing d to $d + 3$. Thus, $r(n + 3) = r(n) + 3$, satisfying the claim of the lemma.

(II)   The last digit d of $\langle n \rangle$ is 7, 8, or 9. In that case we form $\langle n + 3 \rangle$ from the string $\langle n \rangle = d_m d_{m-1} \ldots d_1 d$ ($m \geq 0$) of digits by the following rule, which exhausts three possible subcases:

(1)   If $d_1 \neq 9$, set $\langle n + 3 \rangle = d_m d_{m-1} \ldots (d_1 + 1)(d - 7)$. (If $m = 0$, this rule changes $\langle n \rangle$ to $1(d - 7)$.) Then $r(n + 3) = 1 + r(n) - 7 = r(n) - 6$, satisfying the claim of the lemma.

(2)   If $\langle n \rangle = d_m d_{m-1} \ldots d_{k+2} d_{k+1} 9 \ldots 9d$ with $d_{k+1} \neq 9$ (where $1 \leq k \leq m$), set $\langle n + 3 \rangle = d_m d_{m-1} \ldots d_{k+2}(d_{k+1} + 1)0 \ldots 0(d - 7)$. Then $r(n + 3) = r(n) - 9k - 6$, satisfying the claim of the lemma.

(3)   If $\langle n \rangle = 9 \ldots 9d$, set $\langle n + 3 \rangle = 10 \ldots 0(d - 7)$. Then $r(n + 3) = r(n) - 9m - 6$, satisfying the claim of the lemma.

Having verified the lemma for all subcases, we have proved it to be true.   □

With this lemma we have an immediate *corollary* — a result which follows so easily that it is almost part of the original result:

3 Corollary.

If $n = 3m$, then $r(n)$ is a multiple of 3.
If $n = 3m + 1$, then $r(n)$ is of the form $3k + 1$.
If $n = 3m + 2$, then $r(n)$ is of the form $3k + 2$.

PROOF. The proof is by induction on m for each of the three cases. However, it is so obvious from the lemma that we do not bother to write it out.   □

Finally, we come to the proof of Theorem **1**, which uses the corollary just given.

PROOF OF THEOREM **1**.

1. n is a multiple of $3 \Rightarrow r(n)$ is a multiple of 3. This is immediate from the first clause of the corollary to the lemma.
2. $r(n)$ is a multiple of $3 \Rightarrow n$ is a multiple of 3.

We prove this by proving the contrapositive: n is not a multiple of 3 $\Rightarrow r(n)$ is not a multiple of 3. But there are only two subcases, $n = 3m + 1$

and $n = 3m + 2$, and the truth of these two cases is provided by the second and third clauses of the corollary to the lemma.   □

IMPLICATION AND WELL-FORMED FORMULAS

In our discussion of proof techniques in "everyday mathematics," we introduced two symbols, $\Rightarrow$ and $\Leftrightarrow$:

$A \Rightarrow B$ says that whenever A is true then B must also be true.

$A \Leftrightarrow B$ says that A is true if and only if B is true.

It is clear then that $\Leftrightarrow$, which we call (logical) equivalence is represented by the truth table

p	q	$p \Leftrightarrow q$	
T	T	T	
T	F	F	p if and only if q
F	T	F	
F	F	T	

for to say that $p \Leftrightarrow q$ is true is to say that p and q have the same truth values. (Thus $p \Leftrightarrow q = \neg (p \oplus q)$.) The truth table for $\Rightarrow$, which we call *implication* or the *conditional*, is

p	q	$p \Rightarrow q$	
T	T	T	
T	F	F	if p then q
F	T	T	
F	F	T	

The first two lines are clear: $p \Rightarrow q$ is true for p true only if we also have that q is true. The last two lines are a bit more subtle. For $p \Rightarrow q$ to be true, we have that "if p is true, then q is true." If p is false, then it does not matter whether q is true or false — neither gives us evidence against the claim that the truth of p guarantees the truth of q. And so we say $(F \Rightarrow T) = T$ and $(F \Rightarrow F) = T$. Note that (writing $\bar{p}$ as $\neg p$)

$$p \Rightarrow q = \neg p \vee q.$$

The reason the truth table for $(p \Rightarrow q)$ strikes us as "strange" is that in ordinary English we expect some sort of causal connection between the *antecedent* (p) and the *consequent* (q). For example the following statement looks like a "reasonable" implication

"If you heat the metal to 200°C then the metal will melt"

whether or not it is true. On the other hand,

"If Rome is the capital of Italy then the metal will melt"

feels "strange." But $\Rightarrow$ does not address these subtle interrelations, and only offers a truth value for $(p \Rightarrow q)$ when given the truth values of p and q. Thus we have the following truth values:

Proposition	Truth Value
If $1 + 1 = 2$ then Paris is the capital of France	$(T \Rightarrow T) = T$
If $1 + 1 = 2$ then Rome is the capital of France	$(T \Rightarrow F) = F$
If $1 + 1 \neq 2$ then Paris is the capital of France	$(F \Rightarrow T) = T$
If $1 + 1 \neq 2$ then Rome is the capital of France	$(F \Rightarrow F) = T$

One advantage of the truth table definition of $\Rightarrow$ is that it yields the validity of (If $A \wedge B$ is true, then B is true) in all cases:

A	B	$A \wedge B$	$(A \wedge B) \Rightarrow B$
T	T	T	T
T	F	F	T
F	T	F	T
F	F	F	T

In fact, we might say that this logical requirement *forces* the truth table definition of $\Rightarrow$.

With the truth table definition of $p \Rightarrow q$, we can recapture $\Leftrightarrow$, $\wedge$, and $\vee$ by the formulas (see Exercise 2)

$$p \wedge q = \neg(p \Rightarrow \neg q)$$

$$p \vee q = \neg(\neg p \wedge \neg q) = \neg p \Rightarrow q$$

$$p \Leftrightarrow q = (p \Rightarrow q) \wedge (q \Rightarrow p) = \neg((p \Rightarrow q) \Rightarrow \neg(q \Rightarrow p)).$$

We also have the identity

$$p \Rightarrow q = \neg q \Rightarrow \neg p.$$

This corresponds to our observation that to prove $A \Rightarrow B$ is equivalent to proving its *contrapositive* $\bar{B} \Rightarrow \bar{A}$.

Note that, since we already know that any Boolean function can be built up from $\wedge$, $\vee$ and $\neg$, the above formulas show that any Boolean function can be built up using just $\Rightarrow$ and $\neg$. As an exercise in the inductive definitions of Section 2.1, we give (in **4** and **5** below) an explicit inductive definition of well-formed formulas built up using $\Rightarrow$ and $\neg$, and of how they are interpreted to yield truth values. The *syntax* gives the "grammar" which tells us what strings of symbols are *well-formed*; the *semantics* tells us what a formula *means.* For example, the syntax must tell us that $(p \Rightarrow \neg q)$ is well-formed but that $(p \neg q)$ is not well-formed; while the semantics tells us that if p is interpreted as T and q as F, then $(p \Rightarrow \neg q)$ is interpreted as T.

4 Syntax of Well-Formed Formulas of Propositional Logic. We fix a set $\mathscr{X} = \{p, q, r, \ldots\}$ whose elements are called the *propositional letters* or *variables*. Then we define the set of well-formed formulas of propositional logic with respect to $\mathscr{X}$ to be the set $\mathrm{Prop}(\mathscr{X})$ defined inductively as follows:

Each x in $\mathscr{X}$ belongs to $\mathrm{Prop}(\mathscr{X})$.

If x and y belong to $\mathrm{Prop}(\mathscr{X})$, then so do $\neg x$ and $(x \Rightarrow y)$.

Alternatively, we may define the set $\langle \mathrm{Prop\_wff} \rangle$ of propositional *wffs* (well-formed formulas) by the context-free or BNF grammar:

$$\langle \mathrm{Prop\_wff} \rangle ::= \langle \mathrm{Prop\_var} \rangle \mid \neg \langle \mathrm{Prop\_wff} \rangle \mid$$

$$(\langle \mathrm{Prop\_wff} \rangle \Rightarrow \langle \mathrm{Prop\_wff} \rangle)$$

$$\langle \mathrm{Prop\_var} \rangle ::= p \mid q \mid r \ldots$$

where $\langle \mathrm{Prop\_var} \rangle$ denotes the set of propositional variables. Thus we can build up $(\neg x \Rightarrow \neg y)$ by forming $\neg x$ and $\neg y$ from x and y, and then combining them with $\Rightarrow$. But the above rules cannot build up $x \neg y$, for example.

5 Semantics of Well-Formed Formulas of Propositional Logic. An *interpretation* $\mathscr{I} : \mathscr{X} \to \{T, F\}$ assigns a truth value $\mathscr{I}(x)$ to each propositional variable x in $\mathscr{X}$. We extend $\mathscr{I}$ to a map $\hat{\mathscr{I}} : \mathrm{Prop}(\mathscr{X}) \to \{T, F\}$ inductively as follows:
(i) $\hat{\mathscr{I}}(x) = \mathscr{I}(x)$ for each x in $\mathscr{X}$.
(ii) If $\hat{\mathscr{I}}$ is already defined for x and y in $\mathrm{Prop}(\mathscr{X})$, then we define $\hat{\mathscr{I}}(\neg x)$ and $\hat{\mathscr{I}}((x \Rightarrow y))$ by the truth tables

$\hat{\mathscr{I}}(x)$	$\hat{\mathscr{I}}(\neg x)$	$\hat{\mathscr{I}}(x)$	$\hat{\mathscr{I}}(y)$	$\hat{\mathscr{I}}((x \Rightarrow y))$
T	F	T	T	T
F	T	T	F	F
		F	T	T
		F	F	T

Thus, for example, the interpretation of $(\neg x \Rightarrow \neg y)$ follows its inductive construction, e.g., if $\hat{\mathscr{I}}(x) = T$ and $\hat{\mathscr{I}}(y) = F$, then $\hat{\mathscr{I}}(\neg x) = F$ and $\hat{\mathscr{I}}(\neg y) = T$, and so, finally, $\hat{\mathscr{I}}((\neg x \Rightarrow \neg y)) = T$.

Two classes of propositional formulas are particularly important because of their semantic properties. These classes are the *tautologies* and the *contradictions*. A formula is a tautology if it is true under every interpretation. If it is false under every interpretation, it is a contradiction. The formula $(p \wedge q \Rightarrow p \vee q)$ is a tautology; $(p \wedge \bar{p})$ is an example of a contradiction.

As another example of the relationship between syntax and semantics let us consider arithmetic expressions. Here we start with a set $\mathscr{X} = \{p, q, r, \ldots\}$ of *numerical variables*. We define the set $\text{Exp}(\mathscr{X})$ of arithmetic expressions over $\mathscr{X}$ inductively by

(i) Each x in $\mathscr{X}$ belongs to $\text{Exp}(\mathscr{X})$.

(ii) If x and y belong to $\text{Exp}(\mathscr{X})$,

then so do $-x, (x + y)$ and $(x * y)$.

An interpretation then offers a map $\mathscr{I}$ from $\mathscr{X}$ to some suitable set A, and a set of maps $\delta_- : A \to A, \delta_+ : A \times A \to A$ and $\delta_* : A \times A \to A$. Familiar examples take A to be $\mathbf{Z}$ (integer), $\mathbf{Q}$ (rational numbers), or $\mathbf{R}$ (real numbers). In these cases, δ_-, δ_+, δ_* are the appropriate numerical minus, addition, and multiplication, respectively.

FORMAL PROOFS IN PROPOSITIONAL LOGIC

Readers who have studied Euclidean geometry will remember the basic idea of a formal proof. We start with axioms and then combine them to deduce new statements using rules of inference.

An *axiomatization* of propositional logic will provide us with a rigorous and formal definition of the concepts of *theorem* and *proof* in terms of the concepts of *axiom* and *rule of inference*. There are different possible axiom systems for propositional logic, each of which suits our purpose here equally well. We choose a particularly simple one.

6 An Axiom System for Propositional Logic

Axioms: All the axioms are well-formed formulas of one of the three following forms:

1. $A \Rightarrow (B \Rightarrow A)$

2. $(A \Rightarrow (B \Rightarrow C)) \Rightarrow ((A \Rightarrow B) \Rightarrow (A \Rightarrow C))$

3. $(\neg(\neg A) \Rightarrow A)$

where A, B, and C each stands for an arbitrary well-formed formula (*wff*).

The reader should note that the three axiom schemas are all *tautologies*, i.e., whatever particular *wffs* may be substituted for A, B, and C, and no matter what truth values these *wffs* may have as interpretation, the overall formula has truth value T (see Exercise 1).

Next, the notion of a rule of inference is that if the premises are tautologies (the *wff* schemas above the line in the rules below), then the *wff* below the line must be tautology as well. We have two rules of inference.

Rule of Inference 1:

Modus Ponens

$$\frac{A,\ A \Rightarrow B}{B}$$

i.e., if *wff* A and *wff* $A \Rightarrow B$ have already been proved, deduce *wff* B.

This is certainly a valid rule of inference, since we know from the truth table definition of $\Rightarrow$ that when A evaluates to T and $A \Rightarrow B$ evaluates to T, then B must evaluate to T at the same time, since $(T \Rightarrow F) = F$.

For the next rule of inference, we need the notion of substitution. A_p^B is the *wff* obtained by replacing every occurrence of the propositional variable p in A by the *wff* B. This substitution can be defined inductively as follows:

Basis Step: $p_p^B = B$; and $q_p^B = q$ if q is a propositional variable distinct from p.

Induction Step: $(\neg A)_p^B = \neg(A_p^B)$

$$(A \Rightarrow C)_p^B = (A_p^B \Rightarrow C_p^B).$$

Rule of Inference 2:

Substitution

$$\frac{A}{A_p^B}$$

This is clearly valid, since if the *wff* A is true no matter what truth value is assigned to p, then the *wff* A_p^B obtained by replacing every occurrence of the propositional variable p in A by the *wff* B will be true no matter what truth value B might have.

The notions of *proof* and *theorem* may now be defined for axiomatic theories in general. A list of *wff*'s $A_1, A_2, \ldots, A_n$ is a *proof* of A_n if for every $i \leq n$ either

1. A_i is an axiom, or
2. A_i may be derived from *wffs* occurring earlier in the list by applying one of the rules of inference.

The well-formed formula A_n is said to be a *theorem* if there is a proof of A_n.

An axiomatization of propositional logic is *sound* if it has the properties demonstrated above: Each axiom is a tautology; and the rules of inference will produce a tautology as conclusion when given tautologies as premises. It is clear, then, that in a sound axiomatization of propositional logic, the theorems it produces are verifiable by truth tables. That is, every formally derived theorem in the axiom system is a truth table tautology. We have shown that axiom system **6** for propositional logic is sound. In fact, **6** is also *complete*, in that every truth table tautology can be formally derived as a theorem in it. For a proof of completeness, the reader is referred to a standard text on logic.

As an example of a proof using axiom system **6.**, we have the following proof of the theorem $A \Rightarrow A$, where A is an arbitrary *wff*:

1. $(A \Rightarrow (B \Rightarrow C)) \Rightarrow ((A \Rightarrow B) \Rightarrow (A \Rightarrow C))$ by Axiom 2,

2. $(A \Rightarrow (B \Rightarrow A)) \Rightarrow ((A \Rightarrow B) \Rightarrow (A \Rightarrow A))$ by substituting A for C in (1),

3. $A \Rightarrow (B \Rightarrow A)$ by Axiom 1,

4. $(A \Rightarrow B) \Rightarrow (A \Rightarrow A)$ by Modus Ponens from (3) and (2),

5. $(A \Rightarrow (B \Rightarrow A)) \Rightarrow (A \Rightarrow A)$ by substituting $(B \Rightarrow A)$ for B in (4).

6. $A \Rightarrow A$ by Modus Ponens from (3) and (5).

QUANTIFIERS AND PREDICATE LOGIC

Mathematical logic starts with the basic formalizations of such *propositional connectives* as AND ($\wedge$), OR ($\vee$), and NOT ($\neg$) — which is the subject matter of propositional logic we have just discussed — but then goes on to study *predicates*. Predicates, as we saw in Section 1.2, are "relations in truth value form." Let us recall what this means. An n-ary relation R on a set S is a subset of S^n, the Cartesian product of n copies of S. For example, we can define a ternary ($= 3$-ary) relation on the natural numbers by saying

$$(m, n, r) \in R \Leftrightarrow m = n + r. \tag{1}$$

The predicate P corresponding to a relation $R \subset S^n$ is just the characteristic function of R

$$P(s_1, \ldots, s_n) = \begin{cases} T & \text{if } (s_1, \ldots, s_n) \in R, \\ F & \text{if } (s_1, \ldots, s_n) \notin R. \end{cases}$$

For example, if P corresponds to the R of ($*$) we have

$$P(3, 4, 7) = F, \quad P(5, 3, 2) = T, \quad P(117, 100, 17) = T, \quad \text{etc.}$$

Notice that $P(m, n, r)$ does *not* have a truth value by itself, for we have not specified numerical values for m, n, and r, but once we choose actual numbers for these variables, we get a definite truth value. There exist other ways, besides substitution of values, to get a truth value from a predicate. Consider, for example, that $m + 0 = m$ for every m in $\mathbf{N}$. We can rephrase this using the R of (1), by saying that

There exists an n in $\mathbf{N}$ (namely 0) with the property that for every m in $\mathbf{N}$ we have $P(m, m, n)$. (2)

Time and again in mathematics we fix some set S as the "universe" in which all relations hold and then express the fact that some property holds for every x in S, or that there is at least one x in S for which the property holds. We thus have special notations:

$(\forall x)P$ denotes "for all x in the 'universe' under discussion, the predicate P is true."

The symbol $\forall$ is the A of "All" upside down. $(\forall x)$ is called the *universal* quantifier since we assert the truth of P for all x in the universe.

$(\exists x)P$ denotes "there exists an x in the 'universe' under discussion for which the predicate P is true."

The symbol $\exists$ is the E of "Exists" back to front. $(\exists x)$ is called the *existential* quantifier since we assert there exists at least one x for which P is true.

Using this notation, we see that (2) may be re-expressed in the form

$$(\exists n)(\forall m)(m = m + n)$$

and we see that this has the truth value T. On the other hand, the statement

$$(\forall m)(m = m + m)$$

has the truth value F over the universe $\mathbf{N}$ since $1 \neq 1 + 1$; while

$$(\exists m)(m = m + m)$$

has the truth value T since $0 = 0 + 0$.

Note, too, that the order of quantifiers can determine whether a statement is true or false. Continuing with universe $\mathbf{N}$, we have:

$(\forall m)(\exists n)(m + n \geq m + m)$ has truth value T (having chosen m, just take $n = m$), while

$(\exists n)(\forall m)(m + n \geq m + m)$ has truth value F (having chosen n, note that $m = n + 1$ fails).

We have seen how to use $\Rightarrow$ and $\neg$ to build up the *wffs* (well-formed formulas) in propositional logic. In predicate logic we may use variables (such as m and n), function symbols (such as $+$) and predicate symbols (such as $\geq$), together with the propositional connectives and the quantifiers to build up our well-formed formulas. But the treatment of the formal syntax and semantics of the predicate calculus, as well as the axiom system, is beyond the scope of this book. However, we will give some informal examples of the use of predicate calculus. First, here is a rule of inference:

7 Rule of Inference for Predicate Calculus ($\forall$-elimination):

$$\frac{(\forall x)P(x)}{P(a)}$$

for any individual constant a i.e., if we have as a theorem of our logic that "$P(x)$ is true for any value of x" then we should be able to infer that $P(a)$ is a theorem, no matter what the particular "element of the universe" a refers to.

To see this rule in action, let us return to the classical argument:

8 Example.

$$\frac{\text{All men are mortal}}{\text{Socrates is a man}}$$

$$\therefore \text{ Socrates is mortal}$$

First, we introduce two predicates: "Man(x)" is to be interpreted as "x is a man" while "Mortal(x)" is to be interpreted as "x is mortal."

The two premises in **8** can be expressed as

(a) $\qquad\qquad\qquad (\forall x)(\text{Man}(x) \Rightarrow \text{Mortal}(x))$

(b) $\qquad\qquad\qquad \text{Man(Socrates)}.$

Applying $\forall$-elimination to (a), we infer

(c) $\qquad\qquad\qquad \text{Man(Socrates)} \Rightarrow \text{Mortal(Socrates)}.$

Then, applying Modus Ponens to (b) and (c) we obtain the desired conclusion:

(d) $\qquad\qquad\qquad \text{Mortal(Socrates)}.$

Here is another example:

9 Example.

$$\frac{\text{Any friend of Martin is a friend of John}}{\text{Peter is not John's friend}}$$

$$\therefore \text{ Peter is not Martin's friend}$$

Here we introduce "$F(x, y)$" to be interpreted as "x is a friend of y." The two premises in **9** can then be formalized as:

(a) $\qquad\qquad\qquad (\forall x)(F(x, \text{Martin}) \Rightarrow F(x, \text{John}))$

(b) $\qquad\qquad\qquad \neg F(\text{Peter, John}).$

Applying $\forall$-elimination to (a), we infer

(c) $\qquad\qquad\qquad F(\text{Peter, Martin}) \Rightarrow F(\text{Peter, John}).$

To proceed further, we would invoke (d), which is in fact a theorem of propositional logic:

(d) $\qquad\qquad\qquad (A \Rightarrow B) \Rightarrow (\neg B \Rightarrow \neg A).$

After making the appropriate substitutions of $F(\text{Peter, Martin})$ for A and $F(\text{Peter, John})$ for B, we may apply Modus Ponens to infer

(e) $\qquad\qquad\qquad \neg F(\text{Peter, John}) \Rightarrow \neg F(\text{Peter, Martin}).$

We then apply Modus Ponens to (b) and (e) to obtain the conclusion of our argument

(f) $\qquad\qquad\qquad \neg F(\text{Peter, Martin}).$

EXERCISES FOR SECTION 4.2

1. For each of the following *wffs*, evaluate the truth table to confirm that it is a tautology.
   (i) $p \Rightarrow (q \Rightarrow p)$
   (ii) $(p \Rightarrow (q \Rightarrow r)) \Rightarrow ((p \Rightarrow q) \Rightarrow (p \Rightarrow r))$
   (iii) $(\neg(\neg p) \Rightarrow p)$.

2. Use the truth tables to check the equalities
   (i) $p \wedge q = \neg(p \Rightarrow \neg q)$
   (ii) $p \vee q = \neg p \Rightarrow q$
   (iii) $p \Rightarrow q = \neg q \Rightarrow \neg p$
   (iv) $p \Leftrightarrow q = (p \Rightarrow q) \wedge (q \Rightarrow p)$.

3. Apply the interpretation $\mathscr{I}$ of 5 to evaluate $(\neg(\neg p) \Rightarrow p)$
   (i) when $\mathscr{I}(p) = T$; and
   (ii) when $\mathscr{I}(p) = F$.

4. Translate into *wffs* of predicate calculus
   (i) Not all birds fly.
       Use $B(x)$ for "x is a bird," $F(x)$ for "x can fly."
   (ii) Anyone who is persistent can learn logic.
       Use $P(x)$ for "x is persistent," $L(x)$ for "x can learn logic."
   (iii) Everyone loves somebody and no one loves everybody.
       Use $L(x, y)$ for "x loves y."

5. In the following formulas, we consider variables to range over the natural numbers N, and give $+$, $*$, and $\leq$ their usual interpretations. State the truth value of the proposition, or specify what values of the variables will make the truth value equal T.
   (i) $(\forall x)(\exists y)(x^2 + y^2 \leq 100)$
   (ii) $(\exists y)(\exists x)(x^2 + y^2 \leq 100)$
   (iii) $(\forall x)(x + y = x)$
   (iv) $(\forall x)(\exists y)(x + y = x)$
   (v) $(\forall y)(\forall z)(x * y = x * z)$.

6. A propositional calculus formula A is *satisfiable* if it is true for some interpretation. Thus the formula $p \vee q$ is satisfiable, since, for example, $p \vee q$ is true when p is true and q is false. On the other hand, $p \wedge \bar{p}$ is unsatisfiable. Consider those formulas using implication as the only logical connective. Show that every such formula is satisfiable. (Hint: prove your result by induction on the number of implication symbols in the formula.)

7. Let $<$ be the binary order relation on the set of all integers, $*$ be multiplication between two integers, and $+$ addition between two integers. Determine the truth value of each of the following assertions over the universe $\mathbf{Z}$ of all integers.
   (i) $\forall x \forall y \, [x * y = y * x]$
   (ii) $\forall x \exists y \, [x < y]$
   (iii) $\exists x \forall y \, [x < y]$
   (iv) $\forall x \forall y \, [x < y]$
   (v) $\forall x \forall y \, [(x < y) \vee (x = y) \vee (y < x)]$
   (vi) $\forall x \exists y \, [x * x = y]$
   (vii) $\forall x \exists y \, [x = y * y]$
   (viii) $\forall x \forall y \forall z \, [x * (y + z) = (x * y) + (x * z)]$
   (ix) $\forall x \forall y \, [(x * x < y * y) \Rightarrow (x < y)]$
   (x) $\forall x \forall y \forall z \, [(x + z < y + z) \Rightarrow (x < y)]$

8. Give an example of one-place predicates $P(x)$ and $Q(y)$, where $x, y \in \mathbf{N}$, such that

$$\exists x P(x) \wedge \exists y Q(y) \text{ is true,}$$

   but

$$\exists x [P(x) \wedge Q(x)] \text{ is false.}$$

CHAPTER 5

Binary Relations, Lattices, and Infinity

Section 5.1 introduces the two most popular flavors of binary relations on a set, namely equivalence relations and partial orders. Modular arithmetic is seen as a special case of forming equivalence classes, and we see that a variety of objects of interest to computer scientists can be related to a partial order. Particularly important are those partial orders for which we can speak of maxima and minima. We discuss these objects — called lattices — in Section 5.2. There we pay particular attention to the Boolean algebras, an important class of lattices which include the truth values equipped with the operations of disjunction and conjunction, and the set of subsets of a set equipped with union and intersection. Then, in Section 5.3, we show that some infinities are bigger than others! This may seem irrelevant to computer science, but in fact quite the opposite is true — the technique used in this proof, Cantor's diagonal argument, is one of the most important techniques in the theory of computability, and can be used to demonstrate that certain problems cannot be solved by any computer program. Finally, we conclude this chapter with a general, abstract look at trees. Tree structures have shown up in a number of important contexts so far in this book, and the treatment they are given in Section 5.4 highlights some of the algebraic features these structures have in common.

5.1 Equivalence Relations and Partial Orders

We saw in Section 1.3 that a relation $R: A \to B$ from a set A to a set B could be thought of as a subset R of $A \times B$, so that we could then use

$$aRb \quad \text{and} \quad (a, b) \in R$$

as two different notations for saying that "a is R-related to b." When a relation R holds between elements of the same set, $R: A \to A$, we say that such an R is a *relation on A*. In this section we consider two special kinds of relations of this type.

EQUIVALENCE RELATIONS

Perhaps the most familiar relationship is equality. If we write Δ_A for the *equality* relation on A, we have

1
$$\Delta_A = \{(a, b) | a = b\} \subset A \times A.$$

We draw the graph for Δ_N in Figure 32 — we see it is the *diagonal*. (The symbol Δ does not stand for a triangle here. It is the capital letter "delta" — the Greek D. Thus — D for diagonal.)

Incidentally, we recognize from the definition **1** and from Figure 32 that Δ_A is the *graph* (in the sense of Section 1.3) of the *identity function* on A, $id_A: A \to A, a \mapsto a$.

Here is another example of a relation on pairs from the same set. We say that two integers n and n' are *equivalent modulo m* if their difference is divisible by m, where m is some fixed integer greater than 0.

2
$$n \sim_m n' \Leftrightarrow (n - n') \bmod m = 0$$

i.e., the remainder on dividing $n - n'$ by m is zero. We also write $n \equiv n' \bmod m$ to indicate that $(n - n') \bmod m = 0$, so that we may define the relation $\sim_m$ as a subset of $\mathbf{Z} \times \mathbf{Z}$ by rewriting 2 in the form

$$\sim_m = \{(n, n') | n \equiv n' \bmod m\} \subset \mathbf{Z} \times \mathbf{Z}.$$

We now observe three properties of $\sim_m$ and ask the reader to note that equality enjoys the same three properties.

1. Each element is related to itself (we say the relation is *reflexive*): $n - n = 0$ so certainly $n \sim_m n$ for each n in $\mathbf{Z}$.
2. If two elements are related, the order does not matter (we say the relation is *symmetric*):
   If $(n - n')$ is divisible by m, then certainly $(n' - n)$ is divisible by m as well. Thus $n \sim_m n'$ implies $n' \sim_m n$.
3. If n is related to n', and n' is related to n'', then n is related to n'' (we say the relation is *transitive* because it can make the transition from n to n'' across the common relative n'):

   $n \sim_m n'$ means $n = n' + km$ for some integer k.

   $n' \sim_m n''$ means $n' = n'' + k'm$ for some integer k'.

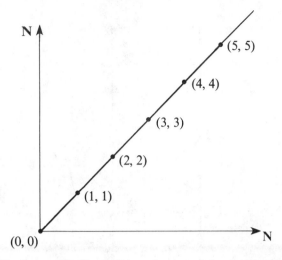

Figure 32   The dots on the diagonal show 6 points of the equality relation Δ_N.

But then

$$n = n' + km = (n'' + k'm) + km = n'' + (k' + k)m$$

and so $n \sim_m n''$ as was to be shown.

This leads us to the following general definition, of which Δ_A and $\sim_m$ are examples:

3 Definition. We say that a relation $\equiv$ on the set A is an *equivalence relation* if it satisfies the three conditions:

R (*Reflexivity*): For every a in A, $a \equiv a$.
S (*Symmetry*): For every a, b in A, $a \equiv b$ implies $b \equiv a$.
T (*Transitivity*): For every a, b, c in A, if $a \equiv b$ and $b \equiv c$ then $a \equiv c$.

Recall now, Figure 31 of Section 4.1, which we reproduce here as Figure 33. This shows the map

$$\mathbf{Z} \to \mathbf{Z}_m, \qquad n \mapsto n \bmod m$$

where $\mathbf{Z}_m = \{0, 1, \ldots, m - 1\}$. We stress that two integers in $\mathbf{Z}$ map to the same entry in $\mathbf{Z}_m$ just in case they are equivalent, and that $\mathbf{Z}$ breaks neatly into nonoverlapping subsets, one for each equivalence class, i.e., collection of elements equivalent under $\sim_m$. The next definition and proposition show us that this "partitioning" will be achieved by any equivalence relation.

4 Definition. Let $\equiv$ be an equivalence relation on the set A. Then for each a in A, *the equivalence class of a with respect to $\equiv$*

$$[a]_\equiv = \{b \mid a \equiv b\}$$

is the set of all elements of A equivalent to a.

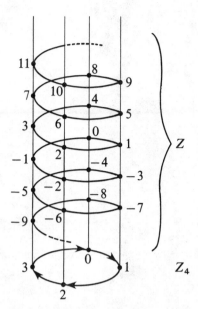

Figure 33   The mapping from $\mathbf{Z}$ to $\mathbf{Z}_m$ for the case $m = 4$. The modulo values appear on the bottom ring. Integers are in the same equivalence class just in case they lie on the same vertical line.

When the context makes clear which equivalence relation we are considering we shall use the simple notation $[a]$, rather than $[a]_\equiv$, to denote the equivalence class of a with respect to $\equiv$.

5 Proposition. *Let* $\equiv$ *be an equivalence relation on the set A. Then for any two elements a and b of A, either* $[a]$ *and* $[b]$ *are disjoint, or they are equal.*

PROOF. Given $[a]$ and $[b]$, either they have no elements in common and so are disjoint; or they have at least one element in common. We must show that the latter case implies that $[a] = [b]$.

Consider, then, the case in which c belongs to $[a] \cap [b]$. We shall prove that we have $[a] \subset [b]$. In other words we must prove that $d \in [a] \Rightarrow d \in [b]$. Now $c \in [a]$ means $a \equiv c$; $c \in [b]$ means $b \equiv c$; and $d \in [a]$ means $a \equiv d$. But then $c \equiv a$ by symmetry; and then ($b \equiv c$ and $c \equiv a$) implies $b \equiv a$ by transitivity; and then ($b \equiv a$ and $a \equiv d$) implies $b \equiv d$ by transitivity. Thus any d in $[a]$ also belongs to $[b]$, so that $[a] \subset [b]$.

It is now immediate by symmetry that $[b] \subset [a]$. But $[a] \subset [b]$ and $[b] \subset [a]$ together imply that $[a] = [b]$, and so our proof is complete. □

Let us return to $\sim_2$ on $\mathbf{Z}$ and write $[n]_2$ for $[n]_{\sim_2}$. Two integers n and n' belong to the same equivalence class just in case they differ by a multiple of 2.

Thus

$$[0]_2 = [-2]_2 = [2]_2 = \cdots = [17364]_2$$
$$= \cdots = \text{the set } \mathbf{Z}_e \text{ of all } even \text{ integers,}$$

$$[1]_2 = [-1]_2 = [3]_2 = \cdots = [2957]_2$$
$$= \cdots = \text{the set } \mathbf{Z}_0 \text{ of all } odd \text{ integers.}$$

Thus there are only two equivalence classes, $\mathbf{Z}_e$ and $\mathbf{Z}_0$, but each of them has infinitely many names. We may pick 0 and 1 as the typical representatives, so that we have

$$\mathbf{Z}/\!\sim_2 \ = \{[0]_2, [1]_2\}$$

where we use the notation $\mathbf{Z}/\!\sim_2$ to denote the set of equivalence classes of $\mathbf{Z}$ under $\sim_2$. We recognize $\mathbf{Z}/\!\sim_2$ as just being a relabeled version of $\mathbf{Z}_2 = \{0, 1\}$. In this light, we may view Figure 33. as showing the map $\mathbf{Z} \to \mathbf{Z}_4$ which sends each integer n to its equivalence class, $[n]_4$, — simply by replacing the labels 0, 1, 2 and 3 by $[0]_4$, $[1]_4$, $[2]_4$ and $[3]_4$, respectively, on the $\mathbf{Z}_4$ circle.

This all leads to the following general definition:

6 Definition. Given an equivalence relation $\equiv$ on the set A, we define "A modulo $\equiv$" to be the set

$$A/\!\equiv \ = \{[a]_\equiv | a \in A\}$$

of all equivalence classes of A. (Note that $A/\!\equiv$ has one element for each *distinct* equivalence class; *not* one element for each a in A.) The map

$$\eta_\equiv : A \to A/\!\equiv, \qquad a \mapsto [a]_\equiv$$

which sends each element to its equivalence class is clearly onto, and is called the *canonical surjection.*

7 Example. Let $\mathbf{Z}_+$ be the set $\{n | n > 0 \text{ and } n \in \mathbf{Z}\}$ of positive integers. Define the relation $\sim$ on $\mathbf{Z} \times \mathbf{Z}_+$ by

$$(p, q) \sim (r, s) \Leftrightarrow ps = rq.$$

Then $\sim$ is an equivalence relation because we have:

Reflexivity: $(p, q) \sim (p, q)$ since $pq = pq$.

Symmetry: $(p, q) \sim (r, s) \Rightarrow (r, s) \sim (p, q)$ since $ps = rq \Rightarrow rq = ps$.

Transitivity: $(p, q) \sim (r, s)$ and $(r, s) \sim (t, u) \Rightarrow (p, q) \sim (t, u)$ since if $ps = rq$ and $ru = ts$ then $(pu)s = (ps)u = (rq)u = (ru)q = (ts)q = (tq)s$. But $s \in \mathbf{Z}_+$ and so we may cancel it to deduce that $pu = tq$, i.e., $(p, q) \sim (t, u)$.

We can thus define the set $\mathbf{Z} \times \mathbf{Z}_+/\sim$ of equivalence classes modulo $\sim$, and discover that it is just (a relabeled version of) the set $\mathbf{Q}$ of rational numbers (i.e., fractions). To see this, use the notation

$$\frac{p}{q} \quad \text{for } [(p, q)]_\sim .$$

Then the definition of $\sim$ says that $[(p, q)]_\sim = [(r, s)]_\sim$ if and only if $(p, q) \sim (r, s)$, i.e., if and only if $ps = rq$. But this just says

$$\frac{p}{q} = \frac{r}{s} \quad \text{if and only if} \quad ps = rq$$

which is indeed the familiar definition of equality of rational numbers.

PARTIAL ORDERS

We now consider a different kind of relation on a set A, namely a partial order. To motivate this, consider the relation $\leq$ of "less than or equal to" on the set $\mathbf{Z}$ of integers. It has the reflexive property — certainly $n \leq n$ for each integer n. It is also transitive: $m \leq n$ and $n \leq p$ certainly implies $m \leq p$ for any three integers m, n, and p. But it is *not* symmetric: for if $m \leq n$ we cannot have $n \leq m$ unless $m = n$. These three properties taken together define a *partial order*, which we summarize formally below.

8 Definition. We say that a relation $\leq$ on the set A is a *partial order* if it satisfies the three conditions:

R (*Reflexivity*):   For every a in A, $a \leq a$.
A (*Antisymmetry*):   For every a, a' in A, if both $a \leq a'$ and $a' \leq a$ hold, then $a = a'$.
T (*Transitivity*):   For every a, a', a'' in A, if both $a \leq a'$ and $a' \leq a''$ then $a \leq a''$.

If $\leq$ is a partial order on A, we call the pair $(A, \leq)$ a *poset* (short for *partially ordered set*).

In fact, "less than or equal to" is a *total order* on A, i.e., a partial order with the extra property that every pair of elements a, a' of A are related — either $a \leq a'$ or $a' \leq a$. However, the next example shows that a partial order need not be total (which is why we call it partial!). If $\leq$ is not a total order on A, we say two elements a, a' are *incomparable* (with respect to $\leq$) if neither $a \leq a'$ nor $a' \leq a$.

9 Example. Given any set S, let 2^S be the set of all subsets of S, and let $\subset$ be the usual inclusion relation

$$A \subset B \Leftrightarrow (s \in A \text{ implies } s \in B).$$

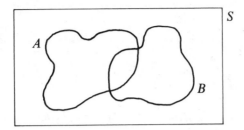

Figure 34   *A* and *B* are incomparable in $(2^S, \subset)$.

Then $(2^S, \subset)$ is a poset: it is clearly reflexive, antisymmetric ($A \subset B$ and $B \subset A$ implies $A = B$), and transitive ($A \subset B$ and $B \subset C$ implies $A \subset C$). But $\subset$ is *not* a total order, since the *A* and *B* of the Venn diagram of Figure 34 are clearly incomparable.

10 Example. We saw in Section 1.3 that a partial function $f: A \to B$ is a rule which assigns to each *a* in some subset dom(f) of *A*, a unique element $f(a)$ in *B*. We thus saw that the graph of such an *f* is a subset of $A \times B$ with the property that if (a, b) and (a, b') belong to that set for the same *a*, we must have $b = b'$. We can thus define a partial order on the set **Pfn**(A, B) of all partial functions from *A* to *B* by

$$f \leq g \Leftrightarrow \text{graph}(f) \subset \text{graph}(g) \text{ as subsets of } A \times B.$$

This says that $f \leq g$ iff "whenever $f(a)$ is defined for a particular *a* in *A*, then $g(a)$ must also be defined with $g(a) = f(a)$," as shown in Figure 35.

We say that two partial functions f_1 and f_2 are *disjoint* if at most one of them is defined at any *a* in A: dom(f_1) $\cap$ dom(f_2) $= \varnothing$. If f_1 and f_2 are disjoint, we may define their *disjoint sum* $f_1 + f_2$ by

$$\text{graph}(f_1 + f_2) = \text{graph}(f_1) \cup \text{graph}(f_2), \text{ the union in } 2^{A \times B}.$$

Figure 35 $f \leq g$: The entire curve is the graph of g, while the darkened portion is the graph of f.

Thus

$$(f_1 + f_2)(a) = \begin{cases} f_1(a) & \text{if } a \in \text{dom}(f_1) \\ f_2(a) & \text{if } a \in \text{dom}(f_2) \\ \text{undefined} & \text{otherwise.} \end{cases}$$

Note that

11 *On* **Pfn**(A, B), $f \le g$ *iff there exists a partial function h disjoint from f such that* $g = f + h$.

PROOF. Suppose $g = f + h$, and f and h are disjoint. Then $f \le g$ since $f \le f + h$. For the converse, let $f \le g$, so that we must have $\text{dom}(f) \subset \text{dom}(g)$. Let then h be defined by $\text{dom}(h) = \text{dom}(g) - \text{dom}(f) = \{a | g(a) \text{ is defined but } f(a) \text{ is not defined}\}$, and then set

$$h(a) = g(a) \quad \text{for each } a \text{ in dom}(h).$$

Then f and h are disjoint, and $g = f + h$, as was to be shown.   □

The relation $\le$ and the partial-addition $+$ on **Pfn**(A, B) play an important role in theoretical computer science in providing the formal setting for *programming language semantics*, the mathematical study of computer programs.

Other important examples of ordering are related to the set X^* of finite strings over the alphabet X.

12 Definition. We say that w is a *prefix* of w', and write $w \le_p w'$ for w, w' in X^*, just in case there is a third string w'' such that $ww'' = w'$.

It is clear that $\le_p$ is a partial order that is not total — for example, $1011 \le_p 1011010$, but the strings 1011 and 1001 are incomparable. We can also place a total order on X^*, the *lexicographic* order, which is the same as the ordering used in making dictionaries ("lexicon" is another word for "dictionary"). First we place a total ordering on X — just as we have to fix the order of the letters in the alphabet before we can order words in the dictionary. Now consider how we tell that APPLE precedes APPRE-CIATE in the dictionary. We remove the longest common prefix and look at the first letter of the remaining suffix — the word comes first whose initial letter comes first. But what of APPLE and APPLESAUCE? We must conclude that Λ precedes SAUCE to make the order work. All of this seems to be captured in the following definition.

13 We define the *lexicographic ordering* $\le_\ell$ on X^*, given a total ordering $\le_e$ on X, as follows, where $x, x' \in X$:

1. $\Lambda \le_\ell w$ for all w in X^*;
2. $x \le_\ell x'$ as strings in X^* if $x \le_e x'$ as elements of X.
3. If $w \le_\ell w'$ and $w \ne \Lambda$ then $uwv \le_\ell uw'v'$ for all $u, v,$ and v' in X^*.

EXERCISES FOR SECTION 5.1

1. Let L be the collection of all lines in the plane.
   (a) Show that the relation

   $$\ell_1 \sim \ell_2 \Leftrightarrow \ell_1 \text{ is parallel to } \ell_2$$

   is an equivalence relation.
   (b) Is the relation

   $$\ell_1 \sim \ell_2 \Leftrightarrow \ell_1 \text{ is perpendicular to } \ell_2$$

   an equivalence relation?

2. Let S be the set of infinite subsets of $\mathbf{N}$, the natural numbers. For A, B in S, say that

   $$A \sim B \Leftrightarrow A \cap B \text{ is infinite.}$$

   Is this an equivalence relation?

3. A relation on the set A is a subset of $A \times A$. We may thus write $R_1 \subset R_2$ to indicate that the subset R_1 is *contained* in the subset R_2, i.e., that $aR_1a' \Rightarrow aR_2a'$. Recall the definition of *composition* of relations from Section 1.3, namely that $aR_2 \cdot R_1a' \Leftrightarrow$ there exists a'' such that aR_1a'' and $a''R_2a'$. We define the *inverse* of the relation R to be $R^{-1} = \{(a, a')|a'Ra\}$. Prove that the relation R on the set A is an equivalence relation if and only if it satisfies the three conditions:
   (i) $\Delta_A \subset R$
   (ii) $R = R^{-1}$
   (iii) $R \cdot R \subset R$.

4. We say a collection of *disjoint* nonempty subsets of A is a partition of A if the union of the sets is A. We know from **Proposition 5** that the equivalence classes of an equivalence relation on A provide a partition of A. Conversely, let us be given any partition of A, i.e., a collection $(A_i|i \in I)$ of subsets of A such that $A_i \cap A_j = \emptyset$ for $i \neq j$, while $\cup(A_i|i \in I) = A$. Define a relation $\sim_A$ on A by $a \sim_A b \Leftrightarrow a$ and b are in the same A_i.
   (i) Prove that $\sim_A$ is an equivalence relation.
   (ii) Prove that the equivalence classes of $\sim_A$ are just the "blocks" A_i of the partition $(A_i|i \in I)$.

5. Suppose x, y, z are strings over some nonempty alphabet X. What familiar binary relation $P(x, y)$ does the following expression of predicate logic represent? (Here, concat is the concatenation function of Section 2.2)

   $$P(x, y) \Leftrightarrow (\exists z)(\text{concat}(x, z) = y).$$

6. Describe informally the meaning of the predicate

   $$Q(x, y) \Leftrightarrow (\exists z)(\text{concat}(z, x) = y)$$

   where x, y, z are strings.

7. Let $(A, \leq)$ be any poset. Define the relation $\geq$ on A by $a \geq a' \Leftrightarrow a' \leq a$. Prove that $(A, \geq)$ is also a poset.

8. Let X^* be the set of finite sequences of elements of the set X, including the empty
    sequence Λ. Show whether or not $(X^*, \leq)$ is a poset with respect to the relations
    defined by
    (i) $w \leq w'$ iff $\ell(w) \leq \ell(w')$, where $\ell(w) = \text{length}(w)$.
    (ii) $w \leq w'$ iff there exist strings w_1, w_2 in X^* such that $w_1 w' w_2 = w$.

9. Let F equal the set of all (total) functions from $\mathbf{N}$ to $\mathbf{N}$. For $f, g \in F$, say that $f \sim g$
    if f and g are equal *almost everywhere*, i.e., if $f(n) = g(n)$ for all but possibly finitely
    many $n \in \mathbf{N}$. Show that $\sim$ is an equivalence relation.

10. Recall the definition of R^*, for any relation $R: A \to A$, from Section 1.3. Prove that
    R^* is reflexive and transitive. Moreover, prove that if $R \subset S: A \to A$, and S is
    transitive and symmetric, then $R^* \subset S$. This is why R^* is called the reflexive transi-
    tive closure of R.

11. This exercise uses the equivalence relation developed in Exercise 9. We denote the
    equivalence class of f by $[f]$. Say that $[f] \leq [g]$ if $f(n) \leq g(n)$ for all but finitely
    many n. (This is numerical, not subset, ordering.)
    (a) Show that $\leq$ is a partial order on these equivalent classes.
    (b) Denote by $[k]$ the equivalence class of $f(n) = k$ for all n. Show that there are in-
        finitely many distinct elements $[f_1], [f_2], \ldots, [f_m], \ldots$, such that

    $$[k] \leq [f_1] \leq [f_2] \leq \cdots \leq [f_m] \leq \cdots \leq [k + 1].$$

12. Let T be a binary tree and define a binary relation, Left, on pairs of nodes of T as
    follows:

    Left$(x, y) \Leftrightarrow x$ and y have a common ancestor z such that
        x is on the left subtree dominated by z and
        y is on the right subtree dominated by z.

    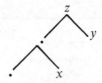

    Show: Left(x, y) and Left $(y, w) \Rightarrow$ Left(x, w).

13. Let $(a_1, a_2, \ldots, a_n)$ be a sequence of all n elements in a finite set A. Suppose R is a
    partial order on A. We say that $(a_1, a_2, \ldots, a_n)$ is a *topological sorting* of A relative
    to R iff for all $a_i, a_j \in A$, $(a_i, a_j) \in R$ implies $i < j$.
    (a) Show that if R is a partial order on a finite set A, then there are elements x,
        $y \in A$ such that for no $z \in A$ is it the case that $(z, x) \in R$ and $(y, z) \in R$. (Such
        elements x and y are respectively called *minimal* and *maximal*.)
    (b) Show that if R is a partial order on a finite set A, then A can be topologically
        sorted relative to R. (Hint: set a_1 equal to some minimal element, throw that
        element out of the partial order, and apply the algorithm again.)

5.2 Lattices and Boolean Algebras

We now look at posets $(A, \leq)$ which have extra properties of interest in theoretical computer science.

First consider $(2^S, \subset)$, the poset of subsets of S ordered under inclusion. Given any family $(A_i | i \in I)$ of sets — whether I is finite or infinite — we may define the union and intersection

$$A_\cup = \bigcup (A_i | i \in I) = \{a | a \in A_i \text{ for at least one } i \in I\}$$

$$A_\cap = \bigcap (A_i | i \in I) = \{a | a \in A_i \text{ for every } i \in I\}.$$

We clearly have $A_i \subset A_\cup$ for every i in I — so we say that $A_\cup$ is an *upper bound* for $(A_i | i \in I)$. Now suppose that the subset B of S is also an upper bound — $A_i \subset B$ for each i. Then

$$a \in A_i \text{ for at least one } i \Rightarrow a \in B$$

and thus $A_\cup \subset B$. In other words, in the inclusion ordering $A_\cup$ is less than or equal to any other upper bound of $(A_i | i \in I)$. We say $A_\cup$ is the *least* upper bound.

1 Definition. Let $(A, \leq)$ be a poset. Let $(a_i | i \in I)$ be a family of elements of A. We say that b is an *upper bound* for this family if $a_i \leq b$ for every $i \in I$. We say that b is the *least upper bound* (*lub*) for the family if $b \leq b'$ for every upper bound b' for $(a_i | i \in I)$; we denote the least upper bound $\bigvee (a_i | i \in I)$. If I has only two elements, we write $a_1 \vee a_2$ (read: a_1 *join* a_2) for $\bigvee (a_1, a_2)$.

We can speak of *the* least upper bound because it is indeed unique. For if b and b' are both least upper bounds we must have

$$b \leq b' \text{ because } b \text{ is least; and}$$

$$b' \leq b \text{ because } b' \text{ is a least upper bound of the family,}$$

and so $b = b'$ by the antisymmetry of $\leq$.

Note immediately that the property of $(2^S, \subset)$ that every family has a least upper bound $\bigvee (A_i | i \in I) = \bigcup (A_i | i \in I)$ is very special.

2 Example. Consider $(\mathbf{N}, \leq)$, the natural numbers ordered by the predicate $\leq$. Clearly any finite sequence of numbers $(n_1, n_2, \ldots, n_k)$ has a least upper bound, namely their maximum: $\bigvee (n_1, n_2, \ldots, n_k) = \max(n_1, n_2, \ldots, n_k)$. For example $\bigvee (3, 4) = 4$, and $\bigvee (2, 1, 3, 1, 16, 4) = 16$.

However, an infinite family of numbers will *not* have an upper bound (unless there are only finitely many different numbers, but with some repeated infinitely often). For example consider the sequence

$$(0, 2, 4, 6, \ldots, 2n, \ldots)$$

of all even numbers. Clearly there is no single number m such that $2n \leq m$ for every n.

3 Example. We may represent many examples of posets $(A, \leq)$ by *Hesse diagrams* such as the two examples shown below. In each case, the nodes represent the elements of A, while the relation is defined by saying that $a \leq a'$ just in case $a = a'$ or there is a path from a to a' which goes entirely in the upward direction. This relation is clearly reflexive, antisymmetric (if two upward paths in sequence return us to the starting point, they must both have zero length) and transitive.

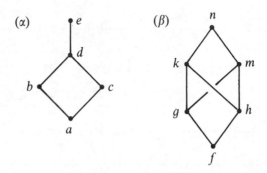

The reader can quickly see that every set of elements in the left-hand poset has a least upper bound — for example, $\bigvee (b, c) = d$ and $\bigvee (c, d) = d$ and $\bigvee (a, b, e) = e$. However, this property fails for the right-hand poset. Just consider the family (g, h) of two elements. Clearly k and m are both upper bounds of g and h which are "as small as possible" but it is not true either that $k \leq m$ or that $m \leq k$, and so there is no least upper bound of (g, h) in the sense of Definition 1.

Now, returning to $(2^S, \subset)$, we see that just as $\bigcup (A_i | i \in I)$ is the least upper bound for the family $(A_i | i \in I)$, so is $\bigcap (A_i | i \in I)$ for greatest lower bound in the sense of the next definition.

4 Definition. Let $(A, \leq)$ be a poset. Let $(a_i | i \in I)$ be a family of elements of A. We say that b is a *lower bound* for this family if $b \leq a_i$ for every $i \in I$. We say that b is the *greatest lower bound* (glb) for the family if $b' \leq b$ for every lower bound b' for $(a_i | i \leq I)$. We denote the greatest lower bound by $\bigwedge (a_i | i \in I)$. If I has only two elements, we write $a_1 \wedge a_2$ (read: a_1 *meet* a_2) for $\bigwedge (a_1, a_2)$.

Thus $\bigwedge (A_i | i \in I) = \bigcap (A_i | i \in I)$ in $(2^S, \subset)$.
For $(\mathbf{N}, \leq)$ we have

$$\bigwedge(n_1, n_2, \ldots, n_k) = \min(n_1, n_2, \ldots, n_k),$$

the smallest integer in the collection. What about infinite collections in $\mathbf{N}$? They *also* have greatest lower bounds! This is because $\mathbf{N}$ is bounded below by 0, and so even infinite collections in $\mathbf{N}$ must have a smallest number ≥ 0. (What about $(\mathbf{Z}, \leq)$?)

In the Hesse diagrams of Example **3** we see that $\wedge$ is well-defined for the poset (α), but $k \wedge m$ is not well-defined in poset (β).

The possession of greatest lower bounds and least upper bounds is so important that posets which possess them are given a special name:

5 Definition. A *lattice* is a poset with the property that every pair of elements of A has both a greatest lower bound and a least upper bound. A *complete lattice* is a poset with the property that every set of elements of A has both a greatest lower bound and a least upper bound.

Thus $(2^S, \subset)$ is a complete lattice for every set S; $(\mathbf{N}, \leq)$ and (α) are lattices, while (β) is not an example of a lattice.

The reader may note that the definition of a lattice only specified the behavior of pairs of elements. In fact, these properties give us the desired properties for any finite set of elements.

6 Proposition. *Let $(A, \leq)$ be a lattice. Then every* finite *nonempty set $(a_1, a_2, \ldots, a_n)$ of elements of A has both a least upper bound and greatest lower bound given, respectively, for $n \geq 2$ by*

$$\bigvee (a_1, a_2, \ldots, a_n) = ((\ldots((a_1 \vee a_2) \vee a_3) \ldots a_{n-1}) \vee a_n)$$

$$\bigwedge (a_1, a_2, \ldots, a_n) = ((\ldots((a_1 \wedge a_2) \wedge a_3) \ldots a_{n-1}) \wedge a_n).$$

PROOF. Note that if $n = 1$, $\bigvee (a_1) = a_1$ and $\bigwedge (a_1) = a_1$ so that every poset trivially has lub's and glb's for one-element subsets. We now turn to the $\bigvee$-formula for $n > 2$. The $\bigwedge$-formula proof is just the same.

Basis Step: For $n = 2$, $\bigvee (a_1, a_2) = a_1 \vee a_2$ by definition.

Induction Step: For $n > 2$, suppose that $\bigvee (a_1, \ldots, a_n) = ((\ldots((a_1 \vee a_2) \vee a_3) \ldots a_{n-1}) \vee a_n)$. We must prove that $\bigvee (a_1, \ldots, a_n, a_{n+1})$ does exist, and is defined by $(\bigvee (a_1, \ldots, a_n)) \vee a_{n+1}$, and we are done.

Let us then set $c = (\bigvee (a_1, \ldots, a_n)) \vee a_{n+1}$. Clearly, by definition, $\bigvee (a_1, \ldots, a_n) \leq c$ — and thus each $a_i \leq c$ for $1 \leq i \leq n$ — and $a_{n+1} \leq c$, and so c is an upper bound for $(a_1, \ldots, a_{n+1})$. Now suppose that d is also an upper bound for $(a_1, \ldots, a_{n+1})$. Then $a_i \leq d$ for $1 \leq i \leq n$, and so $\bigvee (a_1, \ldots, a_n) \leq d$. But also $a_{n+1} \leq d$, and so

$$c = (\bigvee (a_1, \ldots, a_n) \vee a_{n+1}) \leq d$$

and c is the least upper bound for $(a_1, \ldots, a_{n+1})$. $\square$

We often use the notation $(A, \vee, \wedge)$ for a lattice which emphasizes the lattice operations *join*, $\vee$, and *meet*, $\wedge$. This is okay since we can in fact regain the partial order $\leq$ from either lattice operation. The proof is easy, and is left to the reader.

7 Lemma. *Let* $(A, \leq)$ *be a lattice. Then*

1. $a \leq b$ *iff* $a = a \wedge b$.
2. $a \leq b$ *iff* $b = a \vee b$. $\square$

We thus write $(2^S, \cup, \cap)$ and $(\mathbf{N}, \max, \min)$ for our two main examples of a lattice. Here is another interesting example.

8 Example. Let **Boolean** be the set $\{F, T\}$ of truth values, with the ordering $\leq$ given by $T \leq T, F \leq F, F \leq T$. Then $(\mathbf{Boolean}, \leq)$ is indeed a lattice. Using Lemma **7**, we see that the join and meet are given by the tables

p	q	$p \vee q$	$p \wedge q$
T	T	T	T
T	F	T	F
F	T	T	F
F	F	F	F

and we recognize that the join $p \vee q$ is indeed the inclusive-or, p **OR** q, and that the meet $p \wedge q$ is indeed the conjunction, p **AND** q.

Actually, $(\mathbf{Boolean}, \vee, \wedge)$ is the same example as $(2^S, \cup, \cap)$ for the one-element set $S = \{1\}$. To see this, consider the bijection $f \colon \mathbf{Boolean} \to 2^{\{1\}}$ with $f(T) = \{1\}, f(F) = \emptyset$. Then it is easy to see that $\vee$ in **Boolean** is the same as $\cup$ in $2^{\{1\}}$, e.g.,

$$f(T \vee F) = f(T) \cup f(F)$$

since

$$f(T \vee F) = f(T) = \{1\} \quad \text{and} \quad f(T) \cup f(F) = \{1\} \cup \emptyset = \{1\}.$$

Again, $\wedge$ in **Boolean** is the same as $\cap$ in $2^{\{1\}}$, e.g.,

$$f(T \wedge F) = f(T) \cap f(F)$$

since

$$f(T \wedge F) = f(F) = \emptyset \quad \text{and} \quad f(T) \cap f(F) = \{1\} \cap \emptyset = \emptyset.$$

9 Definition. We say the element $\top$ of a poset $(A, \leq)$ is *maximal* if $a \leq \top$ for every a in A, i.e., if $\top = \bigvee(a \,|\, a \in A)$; and that element $\bot$ is *minimal* if $\bot \leq a$ for every a in A, i.e., if $\bot = \bigwedge(a \,|\, a \in A)$.

Thus $\top$ (read: "top") and $\bot$ (which is $\top$ upside down; read: bottom) are unique *if they exist*. In $(2^S, \subset)$ we have $\top = S$, and $\bot = \emptyset$. In $(\mathbf{N}, \leq)$ we have no $\top$, but $\bot = 0$. In $(\mathbf{Boolean}, \leq)$, we have $\bot = F$, and $\top = T$ (corresponding to $\bot = \emptyset$, and $\top = \{1\}$ in $(2^{\{1\}}, \subset)$).

We have shown that the $\wedge$ and $\vee$ of propositional logic may be modeled by the $\cup$ and $\cap$ of set theory. To model negation, we next recall the notion of set complement. For subsets of S we have

$$\bar{A} = S - A = \{a \,|\, a \in S \quad \text{and} \quad a \notin A\}.$$

In the case of the one-element set $S = \{1\}$, this becomes $\bar{\bar{\varnothing}} = \{1\}$ and $\overline{\{1\}} = \varnothing$, which corresponds exactly to negation $p \mapsto \bar{p}$ in **Boolean**: $\bar{F} = T$ and $\bar{T} = F$.

We have the obvious properties

$$A \cap \bar{A} = \varnothing \quad \text{and} \quad A \cup \bar{A} = S$$

in 2^S, which reduce to

$$p \wedge \bar{p} = F \quad \text{and} \quad p \vee \bar{p} = T$$

in **Boolean** on equating F with $\varnothing$ and T with $\{1\}$. This leads to the definition:

10 Definition. We say that a lattice $(A, \vee, \wedge)$ is *complemented* if it has a minimal element $\bot$, a maximal element $\top$, and a map $A \to A$, $a \mapsto \bar{a}$ (we call $\bar{a}$ the complement of a) such that $a \wedge \bar{a} = \bot$ and $a \vee \bar{a} = \top$ for every a in A.

We close by recalling another property of $\cup$ and $\cap$ that was spelled out in the Exercises for Section 1.1, the so-called distributive laws

$$A \cap (B \cup C) = (A \cap B) \cup (A \cap C)$$
$$A \cup (B \cap C) = (A \cup B) \cap (A \cup C).$$

These reduce to the logical equalities

$$p \wedge (q \vee r) = (p \wedge q) \vee (p \wedge r)$$
$$p \vee (q \wedge r) = (p \vee q) \wedge (p \vee r).$$

11 Definition. We say that a lattice $(A, \vee, \wedge)$ is *distributive* if it satisfies the distributive laws

$$a \wedge (b \vee c) = (a \wedge b) \vee (a \wedge c)$$
$$a \vee (b \wedge c) = (a \vee b) \wedge (a \vee c).$$

We may summarize all these observations by saying that each $(2^S, \cup, \cap)$, and so in particular (**Boolean**, $\vee$, $\wedge$), is a Boolean algebra, where:

12 Definition. A *Boolean algebra* is a complemented distributive lattice, i.e., a lattice $(A, \vee, \wedge)$ with the properties:

1. There is a minimal element $\bot$ and a maximal element $\top$ with respect to which every element a in A has a complement $\bar{a}$, i.e., $\bar{a}$ has the property that $a \wedge \bar{a} = \bot$ and $a \vee \bar{a} = \top$.

2 The distributive laws hold:

$$a \wedge (b \vee c) = (a \wedge b) \vee (a \wedge c)$$

$$a \vee (b \wedge c) = (a \vee b) \wedge (a \vee c).$$

It is an amazing fact (due to the American mathematician, Marshall Stone) that *every finite Boolean algebra is isomorphic to* $(2^S, \cup, \cap)$ *for some finite set S*. By this, we mean that if $(A, \vee, \wedge)$ is a Boolean algebra for which A is a finite set, then there is a bijection $f: A \to 2^S$ for some set S which satisfies the property that

$$f(a \vee b) = f(a) \cup f(b) \quad \text{and} \quad f(a \wedge b) = f(a) \cap f(b),$$

just as we saw that the map $T \mapsto \{1\}, F \mapsto \varnothing$ does for that most fundamental of all Boolean algebras, namely (**Boolean**, $\wedge$, $\vee$). Since the proof of the general result takes about five pages, we do not give it here.

EXERCISES FOR SECTION 5.2

1. Show that Definition **5.1.13** makes lexicographic ordering a total order on X^*.

2. Prove that $(X^*, \leq_p)$ has greatest lower bounds for any pair of elements. Does it have least upper bounds?

3. Write out the $\bigwedge$-formula proof for Proposition **6**.

4. Prove Lemma **7**.

5. Verify that under the bijection $f(T) = \{1\}, f(F) = \varnothing$ of Example **8** we do indeed have

$$f(a \vee b) = f(a) \cup f(b)$$

$$f(a \wedge b) = f(a) \cap f(b)$$

for all choices of truth values for a and b.

6. Let $(A, \geq)$ be obtained from $(A, \leq)$ as in Exercise **5.1.7**. Prove that b is a (greatest) lower bound for $(a_i | i \in I)$ in $(A, \leq)$ just in case b is a (least) upper bound for $(a_i | i \in I)$ in $(A, \geq)$.

7. Let $a \to \bar{a}, a \to \tilde{a}$ be two complement operations with respect to which $(A, \vee, \wedge)$ is a Boolean algebra. Prove that $\bar{a} = \tilde{a}$ for every a in A.

5.3 An Introduction to Infinity

Let $\mathscr{F}$ be the collection of all finite sets. Let us write $A \cong B$ to denote that two finite sets A and B are *isomorphic*, i.e., there exists a map $f: A \to B$ with inverse $g: B \to A$, so that $g \cdot f = id_A$ and $f \cdot g = id_B$. We saw in Section 1.3 that $A \cong B$ iff A and B have the same number of elements, $|A| = |B|$. Thus isomorphism of finite sets is an equivalence relation, simply by making use of the fact that equality is an equivalence relation on the natural numbers:

Reflexivity: $|A| = |A|$ for every finite set A.

Symmetry: If $|A| = |B|$ then certainly $|B| = |A|$.

Transitivity: If $|A| = |B|$ and $|B| = |C|$ then $|A| = |C|$.

Now in this section, we want to talk about *infinite* sets. Here, we do not yet know how to talk about the number of elements in an infinite set, except to say that it is "infinite." But is there one infinite number, or two, or as many infinite numbers as there are different infinite sets? Our answer to the question was given by Georg Cantor (1845–1918), who was born in St. Petersburg, Russia, but from 1856 on lived in Germany. Two finite sets are isomorphic if and only if they have the same number of elements. Thus, given two sets A and B, at least one of which is infinite, Cantor *defines* them to have the same number of elements iff they are isomorphic. Let us start following Cantor's approach to infinity by showing that isomorphism is an equivalence relation even for infinite sets. In what follows, we fix some collection $\mathcal{U}$ of sets (the "Universe"), which includes all the finite and infinite sets we need to talk about.

1 Lemma. *Isomorphism, $A \cong B$, is an equivalence relation on $\mathcal{U}$.*

PROOF.

(i)   Reflexivity: $id_A: A \to A$ is an isomorphism, so each set A is isomorphic to itself.

(ii)   Symmetry: If $f: A \to B$ is an isomorphism with inverse $g: B \to A$, then $g: B \to A$ is an isomorphism with inverse $f: A \to B$. Thus $A \cong B$ implies $B \cong A$.

(iii)   Transitivity: Suppose $f: A \to B$ has inverse $g: B \to A$, and $h: B \to C$ has inverse $k: C \to B$. $A \underset{g}{\overset{f}{\rightleftarrows}} B \underset{k}{\overset{h}{\rightleftarrows}} C$. Then $h \cdot f: A \to C$ has inverse $g \cdot k: C \to A$ since

$$(h \cdot f) \cdot (g \cdot k) = h \cdot id_B \cdot k = h \cdot k = id_C$$

and

$$(g \cdot k) \cdot (h \cdot f) = g \cdot id_B \cdot f = g \cdot f = id_A$$

Thus if $A \cong B$ and $B \cong C$, then $A \cong C$.   □

We can thus follow Cantor in making the definition:

2 Definition. We say that two sets A and B in $\mathcal{U}$ *have the same cardinality* if $A \cong B$. If we write $|A|$ for the equivalence class of A under isomorphism, we may write $|A| = |B|$ to show that A and B have the same cardinality. We call each $\cong$-equivalence class a *cardinal number*, and call the cardinal number $|A|$ the *cardinality* of A.

It will cause no trouble to use this notation for finite sets, so that we can think of $|A|$ as being either the integer n, when we wish to count the elements of A, or as the equivalence class of A under isomorphism.[1]

3 Notation. We use the notation $\aleph_0$ (read: aleph-null — aleph is the letter A of the Hebrew alphabet) for the cardinality of the natural numbers:

$$\aleph_0 = |\mathbf{N}|.$$

Our next three examples provide surprises — namely that there are sets much smaller and larger than $\mathbf{N}$ which still have cardinality $\aleph_0$. These examples may tempt us to believe that *every* infinite set has cardinality $\aleph_0$, but we shall see that this is *not* true.

4 Example. Since every positive integer can be paired with a negative integer, we can write down an equation like

5 $$|\mathbf{Z}| + 1 = 2 \times \aleph_0$$

(we have to add 1 to $|\mathbf{Z}|$ because "doubling" $\mathbf{N}$ makes us add a new element, -0 to $\mathbf{Z}$). This might make us expect that $|\mathbf{Z}|$, being "almost twice as big" as $\aleph_0$, cannot equal $\aleph_0$. But, in fact, $|\mathbf{Z}| = \aleph_0$ because the map

$$\mathbf{N} \to \mathbf{Z}, \quad \begin{cases} 2n \mapsto n \\ 2n + 1 \mapsto -(n + 1) \end{cases}$$

is clearly an isomorphism with inverse

$$\mathbf{Z} \to \mathbf{N}, \quad \begin{cases} n \geq 0 \mapsto 2n \\ n < 0 \mapsto 2(-n) - 1. \end{cases}$$

(The reader interested in the "infinite arithmetic" of **5** should work through Exercises 2 and 4 to prove that $\aleph_0 + 1 = \aleph_0 = 2 \times \aleph_0$. It is this type of strange property that makes infinite numbers so very different from finite numbers.)

6 Example. The set of all prime numbers seems much smaller than $\mathbf{N}$. Yet the isomorphism $n \mapsto$ the nth prime number shows that it has cardinality $\aleph_0$, taking 2 as the 0th prime.

---

[1] Indeed some logicians *define* the natural number n to be the equivalence class of all sets isomorphic to $\{0, 1, \ldots, n - 1\}$ — getting the inductive definition going by defining 0 to be $|\varnothing|$, the equivalence class of the empty set. But the reader who finds this confusing — and even good mathematicians have problems the first time they see this! — should "forget about it," because we can just use $|A|$ "either way" for finite sets to achieve the task of this section, which is to shed light on "counting" infinite sets.

7 Example. The set of all *pairs* of natural numbers $\mathbf{N} \times \mathbf{N}$ satisfies the equation

$$|\mathbf{N} \times \mathbf{N}| = \aleph_0 \times \aleph_0 = \aleph_0^2.$$

But we shall see that $\aleph_0^2 = \aleph_0$, again contradicting our intuition from finite numbers, where $n^2 \neq n$ unless n equals 0 or 1. To provide the necessary isomorphism $\mathbf{N} \cong \mathbf{N} \times \mathbf{N}$, we arrange the elements of $\mathbf{N} \times \mathbf{N}$ in a table with (i, j) in row i and column j.

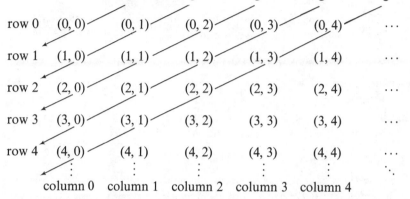

We see that we can count the elements of $\mathbf{N} \times \mathbf{N}$ in the order

$$(0, 0), \ (0, 1), \ (1, 0), \ (0, 2), \ (1, 1), \ (2, 0), \ (0, 3), \ (1, 2), \ldots$$

by moving down the diagonals in turn. Let us use $\langle i, j \rangle$ to denote the position that (i, j) takes in this sequence. Thus

$$\mathbf{N} \times \mathbf{N} \rightarrow \mathbf{N}, \qquad (i, j) \mapsto \langle i, j \rangle$$

is a bijection, and we have $\langle 0, 0 \rangle = 0$, $\langle 0, 1 \rangle = 1$, $\langle 1, 0 \rangle = 2$, etc. Let us see how to write down an explicit formula for $\langle i, j \rangle$. We first note that (i, j) occurs in diagonal $i + j$, and is in fact in position i (row i) in that diagonal. For example, $(1, 3)$ is the 1-position (row 1) in diagonal 4. Thus

$\langle i, j \rangle = $ (the sum of the lengths of diagonals preceding the $(i + j)$th) $+ \ i$

$\qquad = [1 + 2 + \cdots + (i + j)] + i$

$\qquad = \dfrac{(i + j)(i + j + 1)}{2} + i.$

8 Definition. We say a set A is *denumerable* if $|A| = \aleph_0$. An *enumeration* of A is an isomorphism $f \colon \mathbf{N} \rightarrow A$. If we write a_n for $f(n)$, an enumeration is a sequence

$$A = (a_0, a_1, a_2, a_3, \ldots, a_n, \ldots)$$

which lists A without repetitions, and with one element a_n for each natural number n.

The method used in constructing the enumeration of $\mathbf{N} \times \mathbf{N}$ can be formalized in the following lemma, which can be a useful tool for organizing the demonstration that a given infinite set is denumerable.

9 Lemma. *Let A be such that there is an infinite sequence of disjoint nonempty sets $(A_n | n \geq 0)$ whose union is A. If each A_n is finite, then A is denumerable.*

PROOF. Since A_n is finite it has some finite number m_n, say, of elements and so its elements can be written in order as $a_{n1}, a_{n2}, \ldots, a_{nm_n}$. But then

$$a_{01}, \ldots, a_{0m_0}, a_{11}, \ldots, a_{1m_1}, a_{21}, \ldots, a_{2m_2}, \ldots, a_{n1}, \ldots, a_{nm_n}, \ldots$$

is an enumeration of A.   □

In the case of $\mathbf{N} \times \mathbf{N}$, the corresponding A_n is the nth diagonal n, $\{(i, j) | i + j = n\}$, and we enumerate the elements of A_n as

$$(0, n), \ (1, n - 1), \ (2, n - 2), \ldots, (n, 0).$$

We now show that there is a *nondenumerable* infinite set, namely the set $\mathbf{R}$ of all real numbers. The *proof technique* is called *Cantor's Diagonal Argument*, and it turns out to be more important than the theorem! Many important results in mathematics — including theoretical computer science — are proved using variations of this technique, so it is important that you read and re-read the proof to make sure that you understand it fully. [Do exercises 5 and 6.]

10 Theorem. *The set $\mathbf{R}$ of real numbers is nondenumerable.*

PROOF. We must show that there is no isomorphism $\mathbf{N} \to \mathbf{R}$. Since every isomorphism is onto, we can show this by using *Cantor's Diagonal Argument* to prove that there does not exist an *onto* map $f : \mathbf{N} \to \mathbf{R}$. To see this, let f be *any* map $\mathbf{N} \to \mathbf{R}$, which we display by the table

$$
\begin{array}{llllll}
f(0) = a_0. & b_{00} & b_{01} & b_{02} & \cdots & b_{0n} & \cdots \\
f(1) = a_1. & b_{10} & b_{11} & b_{12} & \cdots & b_{1n} & \cdots \\
f(2) = a_2. & b_{20} & b_{21} & b_{22} & \cdots & b_{2n} & \cdots \\
& & \vdots & & & & \\
f(n) = a_n. & b_{n0} & b_{n1} & b_{n2} & \cdots & b_{nn} & \cdots
\end{array}
$$

where each a_n is an integer while b_{nm} is the $(m + 1)$st digit after the decimal point in the decimal expansion of $f(n)$.

Cantor's argument is as follows. Consider the diagonal elements as shown, and let us systematically change each 0 to 1, and each non-zero digit to 0, to form the real number

$$b = 0.\,\bar{b}_{00}\,\bar{b}_{11}\,\bar{b}_{22}\ldots\bar{b}_{nn}\ldots \quad \text{where } \bar{b}_{ij} = \begin{cases} 0 & \text{if } b_{ij} \neq 0, \\ 1 & \text{if } b_{ij} = 0. \end{cases}$$

Thus b certainly belongs to $\mathbf{R}$ but is not in the range $f(\mathbf{N})$ — because b cannot equal $f(n)$ for any n, since the $(n + 1)$st digit of b is $\bar{b}_{nn}$ which does not equal b_{nn}, the $(n + 1)$st digit of $f(n)$. Hence f is *not onto*.

But f was *any* map from $\mathbf{N}$ to $\mathbf{R}$, so there is *no* map of $\mathbf{N}$ onto $\mathbf{R}$.   □

Note that we could add this new element b to the above listing to obtain the listing $g: \mathbf{N} \to \mathbf{R}$ with

$$g(0) = b$$

$$g(1) = f(0)$$

$$\vdots$$

$$g(n) = f(n + 1)$$

$$\vdots$$

But this addition of b does not suffice to make g onto, for we can apply Cantor's argument to the diagonal of g to find a real number not in the range of g — and so on, again and again *ad infinitum*.

EXERCISES FOR SECTION 5.3

1. Let n and n' be two cardinal numbers. Define $\leq$ by saying that $n \leq n'$ just in case $n = |A|$ and $n' = |A'|$ implies that there exists a *one-to-one* map $f: A \to A'$.
   (i) Prove that $\leq$ is well-defined by checking that if $|A| = |B|$ and $|A'| = |B'|$ and $f: A \to A'$ is one-to-one, then there also exists a one-to-one map $g: B \to B'$.
   (ii) Prove that $\leq$ is a partial order on the set $\mathscr{U}/\cong$ of cardinal numbers.

2. Define *addition* of cardinal numbers by

$$|A| + |A'| = |A + A'|$$

where $A + A'$ is the *disjoint union* of A and A'. Define *multiplication* of cardinal numbers by

$$|A| * |A'| = |A \times A'|$$

where $A \times A'$ is the *Cartesian product* of A and A'. Define *exponentiation* of cardinal numbers by

$$|A|^{|A'|} = |A^{A'}|$$

where $A^{A'}$ is the *mapping set* of all functions from A' to A. Prove that each operation is well-defined, i.e., that if $A \cong B$ and $A' \cong B'$ then
   (i) $A + A' \cong B + B'$
   (ii) $A * A \times \cong B * B'$
   (iii) $A^{A'} \cong B^{B'}$.

3. Show that the definition of subtraction

$$|A| - |A'| = |A - A'|$$

where $A - A'$ is the *set difference* $\{a \mid a \in A, a \notin A'\}$ is *not* well-defined by constructing a counterexample using the sets $\mathbf{Z}$ and $\mathbf{N}$.

4. Let n be any natural number. Prove that
   (i) $n * \aleph_0 = \aleph_0$, and
   (ii) $n + \aleph_0 = \aleph_0$.

5. Let n be any natural number ≥ 2. Prove that
   (i) $\aleph_0{}^n = \aleph_0$, and
   (ii) $n^{\aleph_0} \neq \aleph_0$.

6. Let G be the set of all nondecreasing functions $f: \mathbf{N} \to \mathbf{N}$, i.e., each f in G satisfies

$$f(0) \leq f(1) \leq \cdots \leq f(n) \leq f(n + 1) \leq \cdots$$

   Prove that G is *nondenumerable*. [Hint: Modify Cantor's diagonal argument to ensure that the function constructed from the diagonal itself belongs to G.]

7. (i) Let F_n be the set of all maps $f: \mathbf{N} \to \mathbf{N}$ with the properties that (a) $f(k) = 0$ for all $k > n$; and (b) $f(k) \leq n$ for all $0 \leq k \leq n$. How many functions belong to F_n? How many belong to $F_n - F_{n-1}$ (for $n \geq 1$)?
   (ii) Let F be the set of all functions $f: \mathbf{N} \to \mathbf{N}$ with the property that $\{n \mid f(n) \neq 0\}$ is finite. Prove that F is denumerable.

8. We have proved in example 7 that $|\mathbf{N} \times \mathbf{N}| = \aleph_0$. The set $\mathbf{Q}$ of all rational numbers is isomorphic to $\{(m, n) \mid m \in \mathbf{Z}, z \in \mathbf{N} - \{0\}\}/\sim$ where $\sim$ is described in Section 4.1.
   (i) Prove that $|\mathbf{Q}| \leq |\mathbf{N} \times \mathbf{N}|$.
   (ii) Prove that $|\mathbf{N}| \leq |\mathbf{Q}|$.
   (iii) Deduce that $|\mathbf{Q}| = \aleph_0$.

9. Show that $(2^{\aleph_0})^{\aleph_0} = 2^{\aleph_0}$.

10. Determine whether the cardinalities of A and B are equal, in each of the following cases:
    (a) $A = \{n \mid n$ is an integer multiple of $5\}$
        $B = \{n \mid n$ is an integer multiple of $10\}$
    (b) $A = \{2^n \mid n$ is a non-negative integer$\}$
        $B = \{2^{2^n} \mid n$ is a non-negative integer$\}$.

11. Show that the cardinality of the set of all polynomials in one variable with integer coefficients equals $\aleph_0$. What if the coefficients may be drawn from $\mathbf{Q}$, the set of all rational numbers?

12. Show that the set of lines in $\mathbf{R} \times \mathbf{R}$ that go through the point $(0, 0)$ has the same cardinality as $\mathbf{R}$.

13. Let X be a finite set. What is the cardinality of X^*?

5.4 Another Look at Trees

We introduced the concept of an ordered tree in Section 3.2, having already sampled the computer scientist's use of such trees in Section 2.4. We saw, for example, that the s-expression $(a \cdot b) \cdot (a \cdot (b \cdot c))$ could be represented by the tree

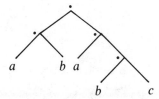

We can also use trees to represent arithmetic expressions

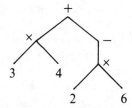

represents $(3 \times 4) + (-(2 \times 6))$

or expressions (such as those in Section 4.2) in propositional logic

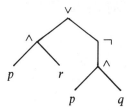

represents $(p \wedge r) \vee (\neg(p \wedge q))$.

In this section we give a different formalization of what counts as a tree structure (from now on all trees will be ordered) and study some of the ways in which trees may be labeled and manipulated.

Perhaps the easiest way to describe a tree is by observing that there is a unique node which acts as the root, and then that each other node can be reached by a path starting at the root. Consider, for example, the node $\bar{v}$

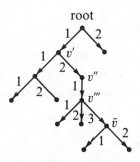

Figure 36

of Figure 36. We can specify it by the sequence $(1, 2, 1, 3)$ — take branch 1 at the root to get v', then take branch 2 at v' to get to v'', then take branch 1 (the only choice) at v'' to get to v''', and finally take branch 3 from v''' to get to $\bar{v}$.

Clearly, then, we can describe each node in the tree by its *address*, namely the string of elements of the set **P** of positive integers describing the branches taken en route from the root. In particular, the root corresponds to the empty string Λ. The question now is: When does a subset of **P*** correspond to the addresses of the nodes of a tree?

Think for a moment of Figure 36 as a family tree showing the women descended from the matriarch represented by the root of the tree. A *successor* of a node corresponds to a *daughter* of the woman represented by that node: if a node has address w in **P***, its successors are the nodes of the tree with addresses wn for some n in **P**. The relation of v corresponding to the *younger sister* of the women represented at v' translates to v and v' being successors of the same node, but with v to the left of v': There is a string w of **P*** and integers m and n with $m < n$ such that the address of v is wm while the address of v' is wn.

Now for Figure 36 to be an acceptable family tree, the following two conditions must be met, where we use T for the set of addresses of nodes of the tree:

1. If a woman is in the family tree, she is either the matriarch, or her mother must also be included in the tree:
   For each w in T, either $w = \Lambda$ or $w = w'n$ with w' in T.
2. If a woman is in the family tree, then either she is the matriarch, or all her younger sisters must also be included in the tree:
   For each w in T, either $w = \Lambda$, or $w = w'n$ with $w'm$ also in T for $1 \leq m \leq n$.

This motivates the following definition:

1 Definition. A subset T of **P*** is called a *tree domain* if:

1. $\Lambda \in T$.
2. If $wn \in T$, with w in **P*** and n in **P** then w and each wm for $1 \leq m \leq n$ also belong to T.

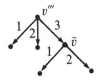

Figure 37

We will often find it convenient to use their addresses as the names of the vertices of a tree. Using this convention, we say:

A node is a *successor* of node w just in case it is wn for some n in **P**.

A node w has *outdegree n* in T — we write $od(w) = n$ — just in case it has n successors, that is wm is in T if and only if $1 \le m \le n$.

A node w is *terminal*, or is a *leaf*, if $od(w) = 0$, i.e., if no wm belongs to T for any m in **P**.

For example, the root of the tree of Figure 36 has 2 successors, while $od(v'') = 1$ and $od(v''') = 3$. There are 7 leaves.

Now consider the node v'''. We may take it as the root of a new tree — the *subtree* of the original tree which is shown in Figure 37. We call the set of addresses of nodes in this subtree the *scope* of v'''. Note that the address of $\bar{v}$ in Figure 36 is 1213, which is 121 (the address of v''' in Figure 36) followed by 3 (the address of $\bar{v}$ in Figure 37). This motivates Definition 2.

2 Definition. Let T be a tree domain, and v a node in T. We say that the *scope* of v in T is the

$$\text{Scope}_T(v) = \{w \mid vw \in T\}$$

which is also a tree domain. We say that a tree domain S is a *subtree* of T just in case it is the scope of some node of S.

OPERATOR DOMAINS

With Figure 38, we recall the examples which introduced the section: In (a) we may label each node of outdegree 2 with the $\cdot$; each leaf with an atom; and there can be no other nodes or labelings.

In (b), we may label each node of outdegree 2 with a $+$ or $*$; each node of outdegree 1 with a $-$; each leaf with an integer; and there can be no other nodes or labelings.

In (c), we may label each node of outdegree 2 with $\vee$ or $\wedge$; each node of outdegree 1 with a $\neg$; each leaf with a p, q or r, or other propositional variable; and there can be no other nodes or labelings.

What is common in each case, then, is that we have a set of labels, and for each label (call it ω) we prescribe a natural number (call it $v(\omega)$ and pronounce it "nu of omega" and call it the "arity" of ω) which tells us the outdegree of the nodes it may label. We call such a collection of labels, each ω

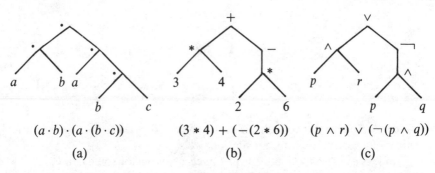

$$(a \cdot b) \cdot (a \cdot (b \cdot c)) \qquad (3 * 4) + (-(2 * 6)) \qquad (p \wedge r) \vee (\neg(p \wedge q))$$

(a)   (b)   (c)

Figure 38

having its own arity $v(\omega)$, an operator domain. In (b), for example, we consider the set of labels $\Omega = \{*, +, -\}$ with $v(*) = v(+) = 2$ and $v(-) = 1$; we then refer to the tree in (b) as an Ω-tree over $\mathbf{Z}$ because it is properly labeled with elements of Ω, but *leaves* may also be labeled by elements of $\mathbf{Z}$.

3 Definition. An *operator domain* is a pair (Ω, v) where Ω is a set, and $v: \Omega \to \mathbf{N}$ is a map which assigns to each *operator label* ω in Ω a natural number $v(\omega)$ called its *arity*. We refer to $\Omega_n = \{\omega \,|\, \omega \in \Omega \text{ with } v(\omega) = n\}$ as the set of n-ary labels.

Let Q be an arbitrary set (which will typically be $\mathbf{N}$ or $\mathbf{Z}$). An Ω-*tree* over Q is a pair (T, h) where T is a tree domain and $h: T \to \Omega \cup Q$ is a labeling function which satisfies the conditions:

$$v \text{ in } T \text{ with } od(v) = n > 0 \Rightarrow h(v) \in \Omega_n$$

$$v \text{ in } T \text{ with } od(v) = 0 \quad \Rightarrow h(v) \in \Omega_0 \text{ or } h(v) \in Q.$$

We can represent tree structures in a linear notation by writing $\omega[t_1, \ldots, t_n]$ rather than

$$\omega$$
$$t_1 \ \cdots \ t_n.$$

Then an Ω-tree over Q may be equivalently defined by induction:

Basis Step: Each ω in Ω_0 and each q in Q is an Ω-tree over Q.
Induction Step: If $t_1, \ldots, t_n$ are Ω-trees over q, and $v(\omega) = n$, then $\omega[t_1, \ldots, t_n]$ is an Ω-tree over Q.

We use $T_\Omega(Q)$ to denote the set of Ω-trees over Q.

We shall now see how this inductive definition allows us to evaluate trees. The present theory generalizes the notion of interpretation of propositional formulas and arithmetic formulas presented in Section 4.2. The

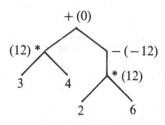

Figure 39

reader may wish to review **4.2.5**, **4.2.6** and the ensuing discussion before proceeding further.

We start with the tree of Figure 38b. Since we know how to manipulate integers, we can evaluate this tree by starting at the leaves and moving rootwards, combining integers with the operators indicated at each node, until we achieve the final value at the root. Figure 39 shows the value attained as we reach each node during this process. The key point is that we have an underlying set **Z** in which all values lie, and each n-ary operator symbol can be interpreted as an n-ary function $\mathbf{Z}^n \to \mathbf{Z}$,,,

$$+ : \mathbf{Z} \times \mathbf{Z} \to \mathbf{Z}, \qquad (m,n) \mapsto m + n$$

$$* : \mathbf{Z} \times \mathbf{Z} \to \mathbf{Z}, \qquad (m, n) \mapsto m * n$$

$$- : \mathbf{Z} \to \mathbf{Z}, \qquad m \mapsto -m.$$

The notion of Ω-algebra — with *carrier* Q replacing the **Z** of this particular example — provides the general setting in which we can evaluate Ω-trees.

4 Definition. Given an operator domain (Ω, v), an Ω-*algebra* is a pair (Q, δ) where Q is a set called the *carrier* and δ assigns to each ω in Ω_n a map

$$\delta_\omega : Q^n \to Q.$$

We have already seen that $(\mathbf{Z}, +, *, -)$ provides an *interpretation* of the $\Omega = \{+, *, -\}$ with $v(+) = v(*) = 2$, $v(-) = 1$. But another interpretation, that is, another Ω-algebra is available among the structures we have studied: $(\mathscr{B}, \delta)$ where $\mathscr{B} = \{T, F\}$, $\delta(+) = \wedge$, $\delta(*) = \vee$, and $\delta(-) = \neg$. It doesn't matter that conjunction is not really addition — all that matters is that the arities match up correctly. Thus the $\mathscr{B}$-tree

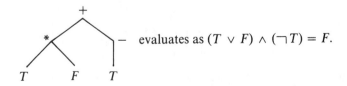

evaluates as $(T \vee F) \wedge (\neg T) = F$.

5 Definition (Inductive Evaluation of Ω-trees over Q). Given an operator domain (Ω, v) and an Ω-algebra (Q, δ), the evaluation by δ of Ω-trees over Q is the map $\delta^*: T_\Omega(Q) \to Q$ given as follows:

Basis Step: The evaluation of q in Q is q; the evaluation of ω with $v(\omega) = 0$ is δ_ω in Q. (An element of Q corresponds to a map $Q^0 \to Q$.)

Induction Step: If $t_1, \ldots, t_n$ have been evaluated as $\delta^*(t_1), \ldots, \delta^*(t_n)$, respectively, and $v(\omega) = n$, then the evaluation of $\omega[t_1, \ldots, t_n]$ is

$$\delta^*(\omega[t_1, \ldots, t_n]) = \delta_\omega[\delta^*(t_1), \ldots, \delta^*(t_n)].$$

The reader should check to see how this compares with Definition **4.2.5**. The assignment of a truth value $\mathscr{I}(x)$ to each propositional variable x in $\mathscr{X}$ generalizes to the assignment of a value δ_ω in Q to each ω with $v(\omega) = 0$. The truth tables for $\neg$ and $\Rightarrow$ may be viewed as maps $\delta_\neg: \mathscr{B} \to \mathscr{B}$, and $\delta_\Rightarrow: \mathscr{B}^2 \to \mathscr{B}$.

POLISH NOTATION AND LEXICOGRAPHIC ORDER

Polish notation is named in honor of the Polish logician Łukasiewicz ("Wukashevitch") who introduced it. Using K for conjunction, A for disjunction and N for negation, he coded propositional logic expressions in the fashion shown below.

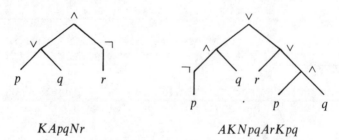

$$KApqNr \qquad\qquad AKNpqArKpq$$

The rule is as follows: To code a tree $\omega[t_1, \ldots, t_n]$ first write down the code for ω, then write down the code for t_1, the code for t_2, etc., and finally write the code for t_n. For example:

Code for $\left(\begin{array}{c}\wedge\\ \vee \quad \neg \\ p \quad q \quad r\end{array}\right) = K\left(\text{code for } \begin{array}{c}\vee\\ p \quad q\end{array}\right)\left(\text{code for } \begin{array}{c}\neg\\ r\end{array}\right)$

$$= KA \text{ (code for } p)(\text{code for } q)N(\text{code for } r)$$
$$= KApqNr$$

where each leaf label is coded as itself.

This suggests the following (if we use each label to code itself):

6 Definition. The *Polish (prefix) notation* of Ω-trees over Q is the map $P: T_\Omega(Q) \to (\Omega \cup Q)^*$ defined inductively as follows:

Basis Step: For each q in Q, $P(q) = q$.
Induction Step: If ω is in Ω_n, and each $t_1, \ldots, t_n$ is in $T_\Omega(Q)$, then $P(\omega[t_1, \ldots, t_n]) = \omega P(t_1) \ldots P(t_n)$.

For example, the Polish notation for

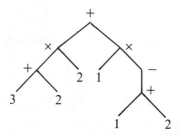

is $+ \times + 3\,2\,2 \times 1 - + 1\,2$. We now tabulate the addresses of the nodes in the order in which their labels occur in this expression.

7

+	×	+	3	2	2	×	1	−	+	1	2
Λ	1	11	111	112	12	2	21	22	221	2211	2212

Think of 1 and 2 as letters of an alphabet, with 1 coming before 2. Then note that the order of words above is lexicographic order — words starting with 1 occur before words starting with 2; and if v is a prefix of $w = vu$, the v comes before w. (Recall definition **5.1.13**.) We close this section by showing that Polish notation always corresponds to lexicographic order. The following rather formal statement corresponds to the situation of table 7 above.

8 Observation. *Let T be a tree domain whose nodes are* $(v_1 = \Lambda, v_2, \ldots, v_k)$ *when arranged in lexicographic order. Let t be the Ω-tree over Q given by the labeling* $h: T \to \Omega \cup Q$ *of T. Then the Polish notation for t is given by*

$$P(t) = h(v_1)h(v_2) \ldots h(v_k),$$

the labels of the nodes arranged in lexicographic order.

PROOF. *Basis Step*: If $t = q$ in Q, then $T = \{\Lambda\}$ and $P(t) = q = h(\Lambda)$. Similarly, if $t = \omega$ with $v(\omega) = 0$.

Induction Step: Let $t = \omega[t_1, \ldots, t_n]$ with ω in Ω_n, $n > 0$, and each t_j in $T_\Omega(Q)$. Let t_j have tree domain T_j with node addresses $(t_{j1}, t_{j2}, \ldots, t_{jn_j})$ in lexicographic order, and assume by way of the inductive hypothesis that

$$P(t_j) = h_j(t_{j1})h_j(t_{j2}) \ldots h_j(t_{jr_j}).$$

Then we see from the picture

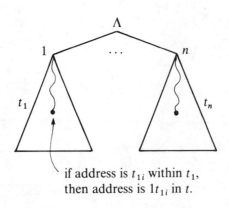

if address is t_{1i} within t_1,
then address is $1t_{1i}$ in t.

that the tree domain T for $t = \omega[t_1, \ldots, t_n]$ is just

$$T = (\Lambda, 1t_{11}, \ldots, 1t_{1r_1}, \ldots, nt_{n1}, \ldots, nt_{nr_n})$$

and that these nodes are in lexicographic order. Moreover, $h(\Lambda) = \omega$ while $h(jt) = h_j(t)$ for each t in T_j. Thus

$$P(t) = \omega P(t_1) \ldots P(t_n)$$

$$= \omega h_1(t_{11}) \ldots h_1(t_{1r_1}) \ldots h_n(t_{n1}) \ldots h_n(t_{nr_n})$$

$$= h(\Lambda)h(1t_{11}) \ldots h(1t_{1r_1}) \ldots h(nt_{n1}) \ldots h(nt_{nr_n})$$

which is precisely the labeling of the nodes of t in lexicographic order.   □

EXERCISES FOR SECTION 5.4

1. We showed that Polish notation for an Ω-tree t over Q corresponds to the lexicographic ordering of the node addresses of t. Show that Polish notation also corresponds to reading the node labels of t when traversed in preorder. (Recall the definition of preorder traversal in Section 3.2.)

2. Establish the proper connections between reverse Polish notation and postorder traversal.

CHAPTER 6

Graphs, Matrices, and Machines

Graph theory started in 1735 with Euler's proof that no path could lead a person over all seven bridges of the Prussian town of Königsberg without at least one bridge being crossed twice. Our study of Euler's theorem in Section 6.1 allows us to introduce such concepts as paths, Euler paths and connectedness for graphs. Section 6.2 uses matrices over semirings to characterize the connectivity of graphs. Finally, Section 6.3 applies these connectivity matrices to analyze the reachability problem for automata, and then returns to a theme initiated in Section 2.3, showing that a language is accepted by a finite-state acceptor iff it can be built up from finite sets by a finite number of applications of the operations of union, dot, and star.

6.1 An Invitation to Graph Theory

DIRECTED AND UNDIRECTED GRAPHS

The word "graph" is used in two senses in theoretical computer science. The first sense — familiar to many readers from analytic geometry or calculus — is that of the graph of a function. As we spelled out in Section 1.3, the graph of a function $f: A \to B$ is the subset

$$\text{graph}(f) = \{(a, b) \mid a \in A, b = f(a)\}$$

of $A \times B$ comprising all pairs related by f. In this section, however, we introduce another sense of graph which is even more widely used in

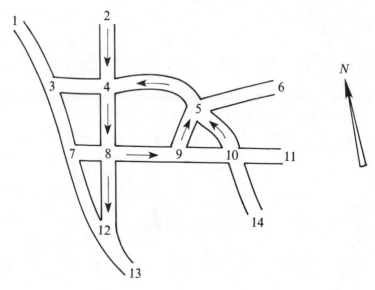

Figure 40

theoretical computer science. Before giving the general definition, we discuss
several examples.

Consider a town map such as that shown in Figure 40, where arrows
indicate one-way streets.

We do not need to keep a detailed record of the shape and geographical
orientation of each street since we can replace the map by the structure
shown in Figure 41a. This object is a *graph* in our new sense. It has one *node*
or *vertex* for each intersection and "town exit" in the original map; and an
edge joining two nodes for each road joining the corresponding points of
the original map. Note that the graph only shows the path relations between
points, not their relative geographical positions. In Figure 41b we show

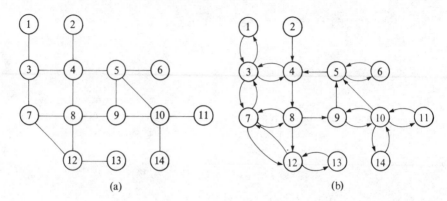

Figure 41

another graph representing the map of Figure 40 — but this times the edges are *directed*, and so we have a single edge for each one-way street, and two oppositely directed edges for each two-way street.

The map of the underground/metro/subway in a major city is more like the graphs of Figure 41 than the map of Figure 41 — but on a subway map each edge is color coded for a different line, so that the number of edges joining two nodes corresponds to the number of different lines which directly link corresponding stations one stop apart.

We may now give the formal definition of a graph in the sense just exemplified.

1 Definition. A *directed graph* $G = (V, E, \partial_0, \partial_1)$ is specified by providing two sets, the set V of *vertices* (or *nodes*) and the set E of *edges*, and two functions, the *start* function $\partial_0: E \to V$ and the *end* function $\partial_1: E \to V$. If $\partial_0(e) = v_0$ and $\partial_1(e) = v_1$ we say that edge *e joins* v_0 to v_1 or *runs from* v_0 to v_1.

2 Definition. An *undirected graph* $G = (V, E, \partial)$ is specified by providing two sets, the set V of *vertices* (or *nodes*) and the set E of *edges*, together with a function ∂ which associates an unordered pair of elements (which need not be distinct) with each edge. If $\partial(e) = \{v_0, v_1\} = \{v_1, v_0\}$ we say that edge *e joins* v_0 to v_1 (and v_1 to v_0).

3 Examples

$$V = \{1, 2, 3\}$$

$$E = \{e_1, e_2, e_3, e_4\}$$

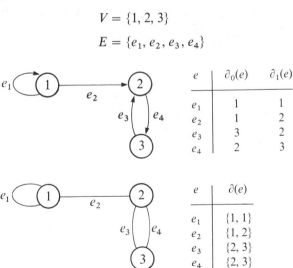

e	$\partial_0(e)$	$\partial_1(e)$
e_1	1	1
e_2	1	2
e_3	3	2
e_4	2	3

e	$\partial(e)$
e_1	$\{1, 1\}$
e_2	$\{1, 2\}$
e_3	$\{2, 3\}$
e_4	$\{2, 3\}$

Note that e_1 in the undirected graph is really associated with a one-element set $\{1\} = \{1, 1\}$, but we readily interpret this as coding the fact that e_1 joins the vertex 1 to itself (it is a *loop*).

The seven bridges of Königsberg

The first interesting result in graph theory (though the subject did not yet have that name) was presented by the Swiss mathematician Leonhard Euler (1707–1783) to the Russian Academy at St. Petersburg (the modern Leningrad) in 1735.[2] Consider the map shown in Figure 42, which represents an island (A) in the middle of the town of Königsberg in Prussia (Germany) around which flow two branches of the river Pregel. Can a person plan a walk in such a way that he will cross each of the seven bridges once and once only (and not swim the river)? You should try to find such a path before reading on.

Euler was told that while some denied that such a walk was possible and others were unsure, no one believed that such a path did exist. But to list all possible sequences of bridge crossings and check that none can include all bridges without repetition takes a lot of work. As a mathematician, Euler asked if there was some simple criterion which would replace the tedious listing of alternatives to answer the following general problem: "Given any configuration of a river and the branches into which it may divide, as well as any number of bridges, determine whether or not it is possible to cross each bridge exactly once." Before giving Euler's answer, and the surprisingly simple proof that it is correct, let us rephrase the problem in the terminology of modern graph theory. First note that the map of Figure 42 can be replaced by the undirected graph of Figure 43 which has a node for each region of land and an edge for each bridge.

We can go from A to D directly via edge (bridge) e, or indirectly via C along the path cg (c followed by g) or dg (d followed by g). An example of a path which crosses all bridges (i.e., involves all the edges) can be written $cgf\,bdgea$ — but, like all such paths across the bridges of Königsberg, it involves a repetition.

4 Definition. A *path* in an undirected graph $G = (V, E, \partial)$ is given by a sequence $a_1, a_2, \ldots, a_n$ of edges to which corresponds a sequence $w_0, w_1, \ldots, w_n$ of vertices such that each edge a_k ($1 \leq k \leq n$) runs from w_{k-1} to w_k. We say that the path runs from (or connects) w_0 to w_n. A path is a *cycle* if $w_0 = w_n$. We say that a cycle is *strict* if no vertex (other than w_0) occurs twice in the sequence.

In Euler's honor, we say that a path $a_1, a_2, \ldots, a_n$ is an *Euler path* if it includes each edge of G once and only once. (Some authors use the term "walk" where we use "path," and then reserve the term path for a walk in

---

[2] An English translation of this presentation appears in Volume 1 (pp. 573–580) of "The World of Mathematics" edited by James R. Newman (Simon and Schuster, 1956). The paper is quick easy reading. Incidentally, "The World of Mathematics" is a delightful collection of essays, excerpts, and even humor which is recommended for browsing and dipping into by anyone who reads this footnote.

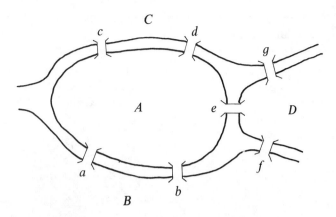

Figure 42   The Seven Bridges of Königsberg

which no edge occurs more than once, but we do not find this distinction useful.)

Before turning to Euler's argument, we use our knowledge of relations to introduce the notion of connectedness of a graph.

5 Definition. Let the relation R be defined on the set V of vertices of the graph G by:

vRv' iff there exists an edge of G which connects v to v'.

Then the reflexive transitive closure $C = R^*$ of R (see **1.3.20**) satisfies the condition

vCv' iff $v = v'$ or there exists a path from v to v'.

6 Observation. *The connectedness relation C for an undirected graph G is an equivalence relation.*

PROOF. Exercise 5.

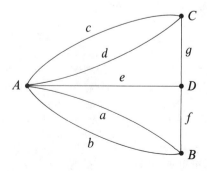

Figure 43   Graph of the Seven Bridges of Königsberg.

Now recall from **5.1.4** the notion of the equivalence class of an element with respect to an equivalence relation.

7 Definition. The *component* of a vertex v of an undirected graph G is the equivalence class of v with respect to the connectedness relation C. We say G is *connected* if $[v] = V$ for any (and then for each, by **5.1.5**) vertex v. We say a vertex v is *isolated* if $[v] = \{v\}$, and that a graph is *proper* if no edge is isolated.

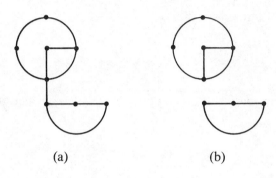

(a)   (b)

Figure 44

The distinct equivalence classes of C are called the *components* of the graph G. Thus a graph is connected iff it has one component.

The graph of Figure 44 (b) is disconnected, and is obtained from the connected graph of (a) by removing one edge. It has two components.

Clearly, a proper graph must be connected if it is to have an Euler path. But this condition is not enough, since the graph of Figure 43 is connected but has no Euler path.

We now use the terminology of Definition **4** to provide the setting for Euler's original argument. Suppose that a graph G has M vertices and N edges. Then an Euler path is a sequence $a_1, a_2, \ldots, a_N$ of the N edges in some appropriate order without repetition, and thus involves $w_0, w_1, w_2, \ldots, w_r$, a corresponding arrangement of the M vertices (with whatever repetitions are necessary). Consider any pair of vertices, say A and B. If the graph G has k edges from A to B, then the pair AB (or BA) must occur as some $w_j w_{j+1}$ in the sequence exactly k times. Moreover, suppose we can write the vertices of G (with repetitions) as a sequence $w_0, w_1, w_2, \ldots, w_n$ where w_j and w_{j+1} are adjacent for $0 \le j < n$. Suppose also that in this sequence each pair of vertices occurs together just as many times as there are edges joining them. Then we can associate distinct edges with each such pair to get an Euler path. Figure 45 (which is also taken from Euler's 1735 article) gives an example of this construction.

8 Euler's Theorem. *Let G be a proper connected undirected graph. Then G has an Euler path iff every node is connected by an even number of edges, or exactly two nodes are connected by an odd number of edges.*

Before we prove the theorem, note that the condition is *not* satisfied by the seven bridges of Königsberg (all 4 vertices of Figure 43 are connected by an odd number of edges), but is satisfied by the graph of Figure 45 (only *D* and *E* are connected by an odd number of edges).

PROOF. *Only if*: Consider a sequence $w_0, w_1, \ldots, w_n$ of vertices that form a path through G. The occurrence of a vertex v as w_0 or w_n requires one edge

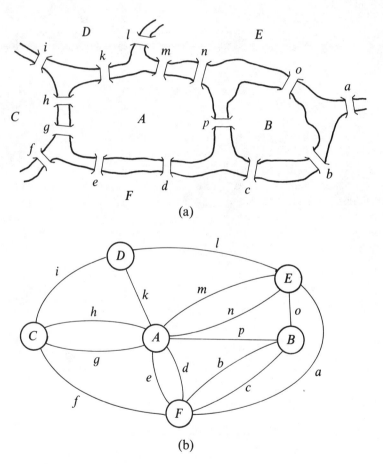

(a)

(b)

$EaFbBcFdAeFfCgAhCiDkAmEnApBoElD.$

(c)

Figure 45   (a) Euler's map. (b) The corresponding graph. (c) An example of an Euler path, with edge labels inserted between the names of the vertices they join.

connected to v, namely, one with $\partial(e) = \{w_0, w_1\}$ or $\partial(e) = \{w_{n-1}, w_n\}$ respectively. The occurrence of v as w_j for $0 < j < n$ requires two edges connected to v, namely e and e' with $\partial(e) = \{w_{j-1}, w_j\}$ and $\partial(e') = \{w_j, w_{j+1}\}$. Thus for an Euler path, either $w_0 = w_n$ and every node is connected by an even number of edges, or $w_0 \neq w_n$ and there are exactly two vertices connected by an odd number of edges.

Interestingly, Euler only proved the "only if" half of his theorem. In fact, the "if" part is harder to prove, and lets us exercise quite a bit of our terminology of graph theory.

If: Let G be a proper connected undirected graph for which all nodes are connected to an even number of edges (the remaining case is considered in Exercise 6.)

The proof proceeds by constructing a sequence of cycles in G which, when taken together, form the desired Euler path. The first cycle, with vertices $w_0, w_1, \ldots, w_n$ and edge sequence $a_1, \ldots, a_n$ is constructed by the following inductive argument.

Basis Step: Pick any vertex v, and set $w_0 = v$.

Induction Step: For $0 \leq j < n$, assume that $w_0, w_1, \ldots, w_j$, and $a_1, \ldots, a_j$ have already been chosen, and that all of a_1 through a_j are distinct. Let E_j be the set of edges not contained in the set $\{a_1, \ldots, a_j\}$.

If $w_j \neq w_0$, then only an odd number of edges connected to w_j have already occurred in the $\{a_1, \ldots, a_j\}$, so at least one edge remains in E_j which is connected to w_j. Let a_{j+1} be such an edge, so that w_{j+1} is the node for which a_{j+1} runs from w_j to w_{j+1}. Finally, set $E_{j+1} = E_j - \{a_{j+1}\}$.

If $w_j = w_0$, either no edges connected to w_j remain, and the cycle is complete, or at least one such edge remains, and the cycle may be further extended as above. (End of Construction.)

Either such a cycle exhausts all edges of the graph and is thus an Euler path, or a number of edges remain. For example, in Figure 46 we might start at node A, and form the cycle

$$AaDcBbA$$

and find no edges to proceed further from A. Note, however, that there are nodes in the above cycle which still have "spare edges." Pick one such node, say B. We then get another cycle:

$$BdDhEiCeDgCfB.$$

Now use this cycle to replace B in the original cycle, and we get

$$AaDcBdDhEiCeDgCfBbA$$

which is indeed an Euler path. In the general case we proceed as follows:

If a cycle is not an Euler path, it must contain at least one vertex connected to edges which do not occur in the cycle. This is obvious if all vertices occur at least once along the cycle. But if any vertex v does not occur on the cycle,

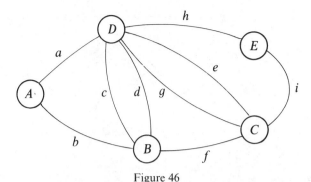

Figure 46

there must be a path from v to any vertex on the cycle (because G is connected), and the desired edge will then be the first edge of that path connected to a vertex on the cycle.

Given a cycle without repeated edges, then, we may extend it by replacing a vertex w by a cycle from v back to itself to get a new longer cycle without repeated edges. Clearly, we can repeat this process only a finite number of times until we have used all the edges of G. The resultant cycle is an Euler path for G. □

MAPS AND FLOWCHARTS

Figure 47 shows how we may associate an undirected graph with a map showing different countries of a region. The *undirected* graph in Figure 47b is obtained from the map of Figure 47a by providing a node for every different country. An edge joins two nodes just in case there is a *stretch* of border (not just a point) where corresponding countries meet.

A famous problem in mathematics is the *Four-Color Problem*. If we look at any map and try to color it in such a way that we never give the same color to countries with a common border, we sometimes need four colors (the reader should check that Figure 47a is an example of a map needing four colors) — but no one had ever found an example of a map needing five colors or more. Could there be a map with a zillion countries that did need five colors? For more than a century, mathematicians tried to prove that no matter how complicated a map might be, four colors would be enough. They simplified the problem by noting that the shape of the countries in the map did not matter, only which pairs of countries had a common border. They could thus simplify the problem to that of placing colors on a graph in such a way that no two nodes with a joining edge have the same color. The key to restriction was that the graph should be *planar* — i.e., that the graph can be drawn on a piece of paper in such a way that no edges cross each other. At first glance, the graph of Figure 47 looks as if it were not planar — but we see that we can rearrange the nodes as shown in Figure 47c to reveal that the

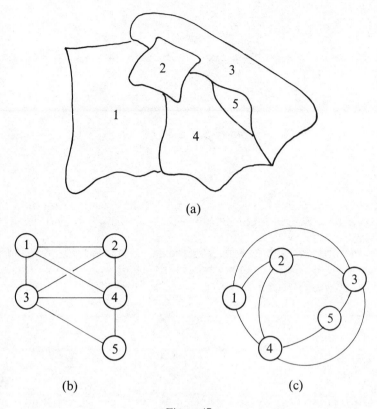

(a)

(b)   (c)

Figure 47

graph really is planar. In fact, the graph of any map must be planar — just position the nodes at the center of each country. Incidentally, this example shows that a graph is given by node-edge relationships, and not by the layout of a picture on a piece of paper — Figures 47b and 47c are two pictures of the same graph. Figure 48 shows a nonplanar graph — note that it requires five colors to color it. Eventually, in 1976, two American mathematicians

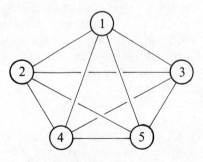

Figure 48

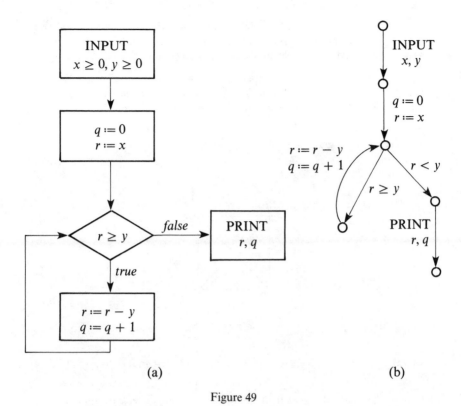

(a)   (b)

Figure 49

were able to prove, with the help of a computer, that in fact any planar graph could be colored using no more than four colors.

Our final example of a graph is the flowgraph of Figure 1 which we used to motivate our discussion of sets and functions. We reproduce it in Figure 49 with one small change — the return arrow of the loop goes to the test box rather than to the line feeding into it. We see that this flow diagram is itself a *directed* graph with one node for each test or operation box, and with a directed edge showing the transfer of control. Every node is labeled, but only the edges leaving a test node bear labels. Figure 49b shows an alternative way of associating a directed graph with a flow diagram. Here we have an edge for each operation, and two edges for each test. All edges are labeled, but none of the nodes are.

EXERCISES FOR SECTION 6.1

1. The *degree* of a vertex i in a graph G is the number of edges incident to the vertex, and we use d_i to note this degree.
   (a) If G has n vertices, prove that $\sum_{1 \le i \le n} d_i$ is an even nonnegative integer.
   (b) Show that every graph has an even number of vertices of odd degree.

2. If G contains at most one edge between any two nodes, what is the length of a shortest cycle? What is the maximum possible length of a strict cycle in G if G has n vertices? What is the maximum possible number of strict cycles in G?

3. A graph is called *bipartite* if its vertices can be partitioned into two sets V_1 and V_2 such that no two vertices in V_1 or in V_2 are joined by an edge; that is, all edges extend "between V_1 and V_2," and no edge is within V_1 or within V_2. Show that G is bipartite iff each of its cycles is of even length.

4. Does the following graph have an Euler path?

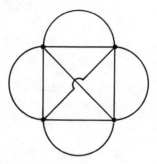

5. Show that the connectedness relation C for an undirected graph G is an equivalence relation.

6. Prove the final case of Euler's theorem: if G is a proper connected undirected graph for which exactly two nodes are connected to an odd number of edges, then G has an Euler path.

6.2 Graphs and Matrices

Most readers will be familiar with matrices from their study of linear algebra. In this section, we shall show how to associate a number of different matrices with a graph, and then use these matrices to deduce various graph properties. But first we recall the basic notions of matrices as used in linear algebra.

MATRICES

We saw in Section 1.1 how the plane could be represented as $\mathbf{R}^2 = \{(x, y) \,|\, x \in \mathbf{R}, y \in \mathbf{R}\}$. More generally, we shall be interested in $\mathbf{R}^n = \{(x_1, x_2, \ldots, x_n) \,|\, x_i \in \mathbf{R}, 1 \le i \le n\}$, the Cartesian product of n copies of $\mathbf{R}$ — these n-tuples may be thought of as the result of n measurements.

Given positive integers m and n, an $m \times n$ matrix A is a collection of $m * n$ real numbers a_{ij} $(1 \le i \le m, 1 \le j \le n)$ arranged in m rows and n columns, with the number a_{ij} (also written $(A)_{ij}$ or A_{ij}) occurring in row i and column j.

$$A = \begin{bmatrix} a_{11} & a_{12} & \cdots & a_{1n} \\ a_{21} & a_{22} & \cdots & a_{2n} \\ & & \vdots & \\ a_{m1} & a_{m2} & \cdots & a_{mn} \end{bmatrix}$$

We may associate with A a map $\mathbf{R}^n \to \mathbf{R}^m$ which sends

$$x = \begin{bmatrix} x_1 \\ x_2 \\ \vdots \\ x_n \end{bmatrix} \text{ in } \mathbf{R}^n \quad \text{to} \quad Ax = \begin{bmatrix} y_1 \\ y_2 \\ \vdots \\ y_m \end{bmatrix} \text{ in } \mathbf{R}^m$$

where the ith component y_i $(1 \le i \le m)$ of Ax is defined as the sum

$$y_i = \sum_{j=1}^{n} a_{ij} x_j = a_{i1} x_1 + a_{i2} x_2 + \cdots + a_{in} x_n.$$

Given two vectors x and x' in $\mathbf{R}^n$ we define their *sum* $x + x'$ in $\mathbf{R}^n$ by setting $(x + x')_i = x_i + x'_i$:

$$\begin{bmatrix} x_1 \\ x_2 \\ \vdots \\ x_n \end{bmatrix} + \begin{bmatrix} x'_1 \\ x'_2 \\ \vdots \\ x'_n \end{bmatrix} = \begin{bmatrix} x_1 + x'_1 \\ x_2 + x'_2 \\ \vdots \\ x_n + x'_n \end{bmatrix}$$

Given a vector x in $\mathbf{R}^n$ and a real number r, the *scalar multiple* rx in $\mathbf{R}^n$ is defined by setting $(rx)_i = rx_i$:

$$r \begin{bmatrix} x_1 \\ x_2 \\ \vdots \\ x_n \end{bmatrix} = \begin{bmatrix} rx_1 \\ rx_2 \\ \vdots \\ rx_n \end{bmatrix}$$

1 Definition. A map $f: \mathbf{R}^n \to \mathbf{R}^m$ is *linear* if for every pair x and x' of vectors in $\mathbf{R}^n$ and pair of real numbers r, r' we have

$$f(rx + r'x') = rf(x) + r'f(x).$$

2 Observation. *A map* $f: \mathbf{R}^n \to \mathbf{R}^m$ *is linear iff it is the map of some* $m \times n$ *matrix* A.

PROOF. (i) Given an $m \times n$ matrix A, we have

$$A(rx + r'x') = \left[\sum_{j=1}^{n} a_{ij}(rx + r'x')_j : 1 \le i \le m \right].$$

But

$$\sum_{j=1}^{n} a_{ij}(rx + r'x')_j = \sum_{j=1}^{n} a_{ij}(rx_j + r'x_j')$$

$$= \sum_{j=1}^{n} (ra_{ij}x_j + r'a_{ij}x_j')$$

$$= \sum_{j=1}^{n} ra_{ij}x_j + \sum_{j=1}^{n} r'a_{ij}x_j'$$

$$= r\sum_{j=1}^{n} a_{ij}x_j + r'\sum_{j=1}^{n} a_{ij}x_j'$$

$$= r \cdot (Ax)_i + r' \cdot (Ax')_i.$$

Thus

$$A(rx + r'x') = r \cdot Ax + r' \cdot Ax'$$

and so the map $\mathbf{R}^n \to \mathbf{R}^m$ induced by A is linear.

(ii)   Let

$$e_1 = \begin{bmatrix} 1 \\ 0 \\ \vdots \\ 0 \end{bmatrix}, \qquad e_2 = \begin{bmatrix} 0 \\ 1 \\ \vdots \\ 0 \end{bmatrix}, \ldots, \qquad e_n = \begin{bmatrix} 0 \\ 0 \\ \vdots \\ 1 \end{bmatrix}$$

be the *n unit vectors* in $\mathbf{R}^n$.

Then any

$$x = \begin{bmatrix} x_1 \\ x_2 \\ \vdots \\ x_n \end{bmatrix}$$

can be written as $x_1e_1 + x_2e_2 + \cdots + x_ne_n$ and it is an easy exercise in linearity to check that if $f: \mathbf{R}^n \to \mathbf{R}^m$ is linear, then

3

$$f(x) = f(x_1e_1 + x_2e_2 + \cdots + x_ne_n) = x_1 f(e_1) + x_2 f(e_2) + \cdots + x_n f(e_n).$$

Now each $f(e_j)$ is a vector in $\mathbf{R}^m$, so there are m real numbers $a_{1j}, a_{2j}, \ldots, a_{mj}$ such that

$$f(e_j) = \begin{bmatrix} a_{1j} \\ a_{2j} \\ \vdots \\ a_{mj} \end{bmatrix}.$$

But then equation **3** becomes:

$$
f(x) = x_1 \begin{bmatrix} a_{11} \\ a_{21} \\ \vdots \\ a_{m1} \end{bmatrix} + x_2 \begin{bmatrix} a_{12} \\ a_{22} \\ \vdots \\ a_{m2} \end{bmatrix} + \cdots + x_n \begin{bmatrix} a_{1n} \\ a_{2n} \\ \vdots \\ a_{mn} \end{bmatrix}
$$

$$
= \begin{bmatrix} a_{11}x_1 + a_{12}x_2 + \cdots + a_{1n}x_n \\ a_{21}x_1 + a_{22}x_2 + \cdots + a_{2n}x_n \\ \vdots \\ a_{m1}x_1 + a_{m2}x_2 + \cdots + a_{mn}x_n \end{bmatrix}
$$

which just says that f is the map corresponding to the matrix

$$
A = \begin{bmatrix} a_{11} & \cdots & a_{1n} \\ a_{21} & \cdots & a_{2n} \\ \vdots & \vdots & \vdots \\ a_{m1} & \cdots & a_{mn} \end{bmatrix} = [f(e_1) \mid f(e_2) \mid \cdots \mid f(e_n)].
$$

$\square$

Given two $m \times n$ matrices A and A' we define their *sum* to be the $m \times n$ matrix $A + A'$ with ij component $a_{ij} + a'_{ij}$. Note that the map of $A + A'$ is the sum of the maps of A and A':

$$
(A + A')x = \left[\sum_{j=1}^{n} (a_{ij} + a'_{ij})x_j \right] = \left[\sum_{j=1}^{n} (a_{ij}x_j + a'_{ij}x_j) \right]
$$

$$
= \left[\sum_{j=1}^{n} a_{ij}x_j + \sum_{j=1}^{n} a'_{ij}x_j \right] = Ax + A'x.
$$

Suppose now that the linear map $f: \mathbf{R}^n \to \mathbf{R}^m$ is given by the $m \times n$ matrix A, and that the linear map $g: \mathbf{R}^m \to \mathbf{R}^p$ is given by the $p \times n$ matrix B. Consider the composite map $g \cdot f: \mathbf{R}^n \to \mathbf{R}^p, x \mapsto g(f(x))$.

$$
g \cdot f(x) = Bf(x) = \left[\sum_{i=1}^{m} b_{ki} f(x)_i \mid 1 \le k \le p \right]
$$

$$
= \left[\sum_{i=1}^{m} b_{ki} \left(\sum_{j=1}^{n} a_{ij} x_j \right) \mid 1 \le k \le p \right] \quad \text{since } f(x) = Ax
$$

$$
= \left[\sum_{j=1}^{n} \left(\sum_{i=1}^{m} b_{ki} a_{ij} \right) x_j \mid 1 \le k \le p \right] \quad \begin{array}{l} \text{reversing the order of} \\ \text{summation} \end{array}
$$

$$
= \left[\sum_{j=1}^{n} c_{kj} x_j \mid 1 \le k \le p \right].
$$

Thus $B(Ax) = Cx$ where the $p \times n$ matrix C is defined by

$$c_{kj} = \sum_{i=1}^{m} b_{ki} a_{ij}.$$

We write $C = BA$, and call C the *product* of B and A (in that order).

MATRICES OVER A SEMIRING

In Section 2.2, we introduced the abstract algebraic notation of a semiring. Recall that a *semiring* is a triple $(S, +, \cdot, 0, 1)$ where

a. $+$ is a commutative associative operation on S with identity 0:

$$a + b = b + a$$

$$(a + b) + c = a + (b + c)$$

$$a + 0 = a = 0 + a$$

for every a, b, c in S.

b. $\cdot$ is an associative operation on S with identity 1:

$$(a \cdot b) \cdot c = a \cdot (b \cdot c)$$

$$a \cdot 1 = a = 1 \cdot a$$

for every a, b, c in S.

c. $\cdot$ is distributive over $+$:

$$a \cdot (b + c) = a \cdot b + a \cdot c$$

for every $a, b,$ and c in S.

To use matrices in studying graphs, we shall consider matrices whose entries come from one of these semirings:

4 $(\mathbf{N}, +, \cdot, 0, 1)$ is the *semiring of natural numbers*, with $+$ (numerical addition with identity 0), and $\cdot$ (numerical multiplication with identity 1).

5 $(\{0, 1\}, \vee, \wedge, 0, 1)$ is the *Boolean semiring* with $\vee$ (disjunction with identity 0), and $\wedge$ (conjunction with identity 1).

6 $(2^{X^*}, \cup, \cdot, \varnothing, \{\Lambda\})$ is the *set of strings semiring* with $\cup$ (union with identity $\varnothing$), and $\cdot$ (concatenation with identity $\{\Lambda\}$).

In what follows, then, we consider **N**-matrices whose entries are natural numbers, Boolean matrices whose entries are 0 or 1, and X^*-matrices whose entries are sets of strings on the alphabet X. In each case we use the semiring operations to define the matrix operations. In the first case, we use the usual addition and multiplication of integers in adding and multiplying matrices:

7 For A and A' being $m \times n$ matrices of non-negative integers

$$(A + A')_{ij} = (a_{ij} + a'_{ij}).$$

For A an $m \times n$, and B a $p \times m$, matrix of non-negative integers

$$(BA)_{kj} = \sum_{i=1}^{m} b_{ki} a_{ij}.$$

For Boolean matrices, we use disjunction ($\vee$) for addition, and conjunction ($\wedge$) for multiplication.

8 A and A' being $m \times n$ matrices of 0's and 1's

$$(A + A')_{ij} = (a_{ij} \vee a'_{ij}).$$

For A an $m \times n$, and B a $p \times m$, matrix of 0's and 1's,

$$(BA)_{kj} = \bigvee_{i=1}^{m} (b_{ki} \wedge a_{ij})$$

the disjunction of m terms, each of which is a conjunction of two terms.

For X^*-matrices, we use union ($\cup$) for addition, and concatenation ($\cdot$) for multiplication.

9 For A and A' being $m \times n$ matrices of subsets of X^*

$$(A + A')_{ij} = (a_{ij} \cup a'_{ij}) = \{w \mid w \in X^*, w \in a_{ij} \text{ or } w \in a'_{ij}\}.$$

For A an $m \times n$, and B a $p \times m$, matrix of subsets of X^*,

$$(BA)_{kj} = \bigcup_{i=1}^{m} (b_{ki} \cdot a_{ij}) = \{w \mid w = w_1 w_2 \in X^* \text{ with } w_1 \in b_{ki} \text{ and } w_2 \in a_{ij}$$

$$\text{for the same } i, 1 \leq i \leq m\}.$$

Returning to maps $\mathbf{R}^n \to \mathbf{R}^m$: Recall that the identity map on any set S is the map $id_S: S \to S$ which sends each s in S to itself, $id_S(s) = s$. In particular, the identity map on $\mathbf{R}^n$ is given by the $n \times n$ *identity matrix*

$$I_n = \begin{bmatrix} 1 & 0 & \cdots & 0 \\ 0 & 1 & \cdots & 0 \\ \vdots & \vdots & & \vdots \\ 0 & 0 & \cdots & 1 \end{bmatrix}$$

whose ij entry is 1 if $i = j$ but is otherwise 0.

The identity matrix I_n has the property that for every $n \times m$ matrix A and every $m \times n$ matrix B

$$I_n \cdot A = A$$

$$B \cdot I_n = B.$$

In fact it is the only matrix with these properties — if J were another $n \times n$ identity matrix, we would have $IJ = J$ because I is the identity, and $IJ = I$ because J is an identity, implying that $I = IJ = J$. (See Exercise 1.)

We now write down the identity matrices for our three new families of matrices. (The proof that they are all identity matrices is left to the reader as Exercise 1(ii). Since the readers need only prove one result to get all three here, they may get some feel for the advantages of using the general notion of a semiring.)

10 The $n \times n$ N-matrix identity is

$$I_n = \begin{bmatrix} 1 & 0 & \cdots & 0 \\ 0 & 1 & \cdots & 0 \\ \vdots & \vdots & \vdots & \vdots \\ 0 & 0 & \cdots & 1 \end{bmatrix}$$

whose ij entry is 1 if $i = j$, but is otherwise 0.

11 The $n \times n$ Boolean matrix identity is the I_n given by

$$\begin{bmatrix} 1 & 0 & \cdots & 0 \\ 0 & 1 & \cdots & 0 \\ \vdots & \vdots & \vdots & \vdots \\ 0 & 0 & \cdots & 1 \end{bmatrix}$$

whose ij entry is 1 if $i = j$, but is otherwise 0.

12 The $n \times n$ X^*-matrix identity is the I_n given by

$$\begin{bmatrix} \{\Lambda\} & \varnothing & \cdots & \varnothing \\ \varnothing & \{\Lambda\} & \cdots & \varnothing \\ \vdots & \vdots & \vdots & \vdots \\ \varnothing & \varnothing & \cdots & \{\Lambda\} \end{bmatrix}$$

whose ij entry is $\{\Lambda\}$, the set whose only element is the empty string Λ in X^*, if $i = j$; but otherwise it is the empty set $\varnothing$. To see why this is the identity matrix, recall that $\varnothing \cdot A = \varnothing$ and $\{\Lambda\} \cdot A = A$ for any $A \subset X^*$.

13 Given any $n \times n$ matrix A, we define its powers A^m for $m \geq 0$ inductively by the formulas:

$$A^0 = I_n, \quad \text{the appropriate } n \times n \text{ identity matrix.}$$

$$A^{m+1} = A^m \cdot A, \quad \text{for each } m \geq 0.$$

Thus $A^1 = I_n \cdot A = A$, $A^2 = A^1 \cdot A = A \cdot A$; $A^3 = A^2 \cdot A = A \cdot A \cdot A$, and A^m for $m > 0$ is the product of m copies of A as desired.

CONNECTION MATRICES OF GRAPHS

Let $G = (V, E, \partial_0, \partial_1)$ be a directed graph with m vertices, and let us write $V = \{v_1, v_2, \ldots, v_m\}$ for ease of reference. Then we associate three connection matrices with G as follows:

14 The *Boolean connection matrix* C of G is the $m \times m$ array of 0's and 1's such that

$$c_{ij} = \begin{cases} 1 & \text{if there is an edge } e \text{ with } \partial_0(e) = v_i, \ \partial_1(e) = v_j \\ 0 & \text{if no edge joins } v_i \text{ to } v_j. \end{cases}$$

15 The **N***-connection matrix* D of G is the $m \times m$ array of natural numbers such that

$$d_{ij} = \text{the number of different edges } e \text{ in } E \text{ with } \partial_0(e) = v_i \text{ and } \partial_1(e) = v_j.$$

16 The *E*-connection matrix* K of G is the $m \times m$ array of subsets of E^* such that

$$k_{ij} = \{e \mid e \in E \text{ with } \partial_0(e) = v_i \text{ and } \partial_1(e) = v_j\}.$$

17 Example. For the directed graph of Example 6.1.3

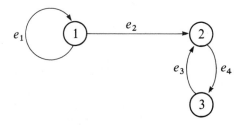

we have

$$C = \begin{bmatrix} 1 & 1 & 0 \\ 0 & 0 & 1 \\ 0 & 1 & 0 \end{bmatrix}, \quad D = \begin{bmatrix} 1 & 1 & 0 \\ 0 & 0 & 1 \\ 0 & 1 & 0 \end{bmatrix}, \quad K = \begin{bmatrix} \{e_1\} & \{e_2\} & \varnothing \\ \varnothing & \varnothing & \{e_4\} \\ \varnothing & \{e_3\} & \varnothing \end{bmatrix}$$

The definitions for an undirected graph $G = (V, E, \partial)$ are essentially the same, except that the phrase "edge e with $\partial_0(e) = v_i$ and $\partial_1(e) = v_j$" simply becomes "edge e with $\partial(v) = \{v_i, v_j\}$" which views the vertices v_i and v_j as forming an *unordered* pair.

18 Example. For the undirected graph of Example 6.1.3

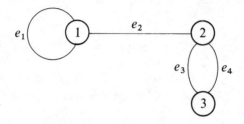

we have

$$C = \begin{bmatrix} 1 & 1 & 0 \\ 1 & 0 & 1 \\ 0 & 1 & 0 \end{bmatrix}, \quad D = \begin{bmatrix} 1 & 1 & 0 \\ 1 & 0 & 2 \\ 0 & 2 & 0 \end{bmatrix}, \quad K = \begin{bmatrix} \{e_1\} & \{e_2\} & \varnothing \\ \{e_2\} & \varnothing & \{e_3, e_4\} \\ \varnothing & \{e_3, e_4\} & \varnothing \end{bmatrix}.$$

Note that for an undirected graph, all these matrices are *symmetric* — the ij element equals the ji element for each pair (i, j).

In what follows, we shall just develop the theory for directed graphs — with Example **18** before them, readers should have no trouble extending the theory to handle the case of undirected graphs.

We begin by defining the notion of a *path* for directed graphs, which parallels the notion of path for an undirected graph given in definition **6.1.4**.

19 Definition. A *path of length n* in the directed graph $G = (V, E, \partial_0, \partial_1)$ from vertex v_i to vertex v_j is a string of length n from E^*

$$e_{i_1} e_{i_2} \cdots e_{i_n}$$

with the property that $\partial_0(e_{i_1}) = v_i$; $\partial_1(e_{i_k}) = \partial_0(e_{i_{k+1}})$ for $1 \le k \le n$; and $\partial_1(e_{i_n}) = v_j$. For each vertex v, we regard the empty string Λ in E^* as being a path from v to itself.

We now examine the powers C^n, D^n, K^n of the connection matrices of our directed graph in Example **17**. By **13**, the 0 powers are the identity matrices:

$$C^0 = \begin{bmatrix} 1 & 0 & 0 \\ 0 & 1 & 0 \\ 0 & 0 & 1 \end{bmatrix}, \quad D^0 = \begin{bmatrix} 1 & 0 & 0 \\ 0 & 1 & 0 \\ 0 & 0 & 1 \end{bmatrix}, \quad K^0 = \begin{bmatrix} \{\Lambda\} & \varnothing & \varnothing \\ \varnothing & \{\Lambda\} & \varnothing \\ \varnothing & \varnothing & \{\Lambda\} \end{bmatrix}.$$

We quickly see that these matrices describe the length 0 paths — K_{ij}^0 is just $\{\Lambda\}$ if $i = j$ and is otherwise empty, reflecting that we cannot get from one node in G to another without traversing at least one edge. D_{ij}^0 is just the number of elements in K_{ij}^0; and C_{ij}^0 is 1 if K_{ij}^0 is nonempty while it is 0 otherwise.

Now compute the squares of the matrices; C, D, and K:

To compute C^2, remember that we use $\wedge$ for multiplication and $\vee$ for addition:

$$C^2 = \begin{bmatrix} 1 & 1 & 0 \\ 0 & 0 & 1 \\ 0 & 1 & 0 \end{bmatrix} \begin{bmatrix} 1 & 1 & 0 \\ 0 & 0 & 1 \\ 0 & 1 & 0 \end{bmatrix} = \begin{bmatrix} 1 & 1 & 1 \\ 0 & 1 & 0 \\ 0 & 0 & 1 \end{bmatrix}.$$

The reader should check that C_{ij}^2 is 1 if and only if there is a path of length 2 from v_i to v_j in our graph.

To compute D^2, we use normal numerical multiplication and addition:

$$D^2 = \begin{bmatrix} 1 & 1 & 0 \\ 0 & 0 & 1 \\ 0 & 1 & 0 \end{bmatrix} \begin{bmatrix} 1 & 1 & 0 \\ 0 & 0 & 1 \\ 0 & 1 & 0 \end{bmatrix} = \begin{bmatrix} 1 & 1 & 1 \\ 0 & 1 & 0 \\ 0 & 0 & 1 \end{bmatrix}$$

and D_{ij}^2 is the number of paths of length 2 from v_i to v_j.

Finally, we compute K^2 and K^3 — where now we use union of subsets of E^* for addition, and concatenation of these sets for multiplication.

$$K^2 = \begin{bmatrix} \{e_1\} & \{e_2\} & \varnothing \\ \varnothing & \varnothing & \{e_4\} \\ \varnothing & \{e_3\} & \varnothing \end{bmatrix} \begin{bmatrix} \{e_1\} & \{e_2\} & \varnothing \\ \varnothing & \varnothing & \{e_4\} \\ \varnothing & \{e_3\} & \varnothing \end{bmatrix} = \begin{bmatrix} \{e_1 e_1\} & \{e_1 e_2\} & \{e_2 e_4\} \\ \varnothing & \{e_4 e_3\} & \varnothing \\ \varnothing & \varnothing & \{e_3 e_4\} \end{bmatrix}$$

while

$$K^3 = K^2 \cdot K = \begin{bmatrix} \{e_1 e_1\} & \{e_1 e_2\} & \{e_2 e_4\} \\ \varnothing & \{e_4 e_3\} & \varnothing \\ \varnothing & \varnothing & \{e_3 e_4\} \end{bmatrix} \begin{bmatrix} \{e_1\} & \{e_2\} & \varnothing \\ \varnothing & \varnothing & \{e_4\} \\ \varnothing & \{e_3\} & \varnothing \end{bmatrix}$$

$$= \begin{bmatrix} \{e_1 e_1 e_1\} & \{e_1 e_1 e_2, e_2 e_4 e_3\} & \{e_1 e_2 e_4\} \\ \varnothing & \varnothing & \{e_4 e_3 e_4\} \\ \varnothing & \{e_3 e_4 e_3\} & \varnothing \end{bmatrix}.$$

We see that K_{ij}^2 is just the set of length 2 paths from i to j in G; while K_{ij}^3 is just the set of length 3 paths from i to j in G.

This calls for a general theorem:

20 Theorem. Let $G = (V, E, \partial_0, \partial_1)$ be a directed graph with E^*-connection matrix K. Then the ij entry of the nth power of K, for any $n \geq 0$, is the set of all length n paths from v_i to v_j in G.

PROOF. We proceed by induction on the power n:

Basis Step:

$$K_{ij}^0 = \begin{cases} \{\Lambda\} & \text{if } i = j \\ \varnothing & \text{if } i \neq j \end{cases}$$

and so is certainly the set of all length 0 paths from v_i to v_j in G.

Induction Step: We verify that truth of the assertion for any n guarantees its truth for $n + 1$:

Suppose then that n is such that $K_{k\ell}^n$ equals the set of all length n paths from v_k to v_ℓ, no matter which vertices v_k and v_ℓ of G we consider. Now consider a path of length $n + 1$ from i to j:

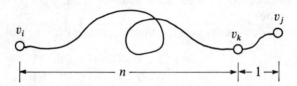

It can be split into a length n path followed by a length 1 path. Thus the set of length $n + 1$ paths from v_i to v_j

$= \bigcup_{v_k \in V}$ {those paths from v_i to v_j consisting of a length n path from v_i to v_k followed by a length 1 path from v_k to v_j}

$= \bigcup_{v_k \in V}$ {length n paths from v_i to v_k} $\cdot$ {length 1 path from v_k to v_j}

$= \bigcup_{v_k \in V} K_{ik}^n \cdot K_{kj}$ by the induction hypothesis and by the definition of K, respectively

$= K_{ij}^{n+1}$ by the definition of multiplication of E^*-matrices.

Thus we have completed the induction step, and the theorem follows. $\square$

We can repeat the same proof with minor modifications to deduce that

21 Theorem. *Let G be a directed graph with Boolean connection matrix C and N-connection matrix D. Then C_{ij}^n equals 1 if and only if there is a length n path from v_i to v_j in G; while D_{ij}^n equals the number of length n paths from v_i to v_j in G.* $\square$

We close by making a simple observation about the path structure of graphs, and then relate it to computations based on the simplest of the connection matrices, namely the Boolean matrix C.

22 Observation. *Let G be a directed graph with m vertices. Then there is a path from v_i to v_j if and only if there is such a path of length $m - 1$ or less.*

PROOF. Obviously, all we have to prove is that if there is a path from v_i to v_j of length greater than or equal to m, there must also be a path of length at most $m - 1$. Now consider a path of length $n \geq m$:

It must visit $n + 1$ nodes — v_i and the end, $\partial_1(e_r)$, of each edge e_r for $1 \leq r \leq n$ where $e_1, \ldots, e_r, \ldots, e_n$ are the edges that make up the path from v_i to v_j. But there are only m different nodes in G so at least one, call it v, must be visited twice.

Thus for some k, $1 \leq k < n$, $\partial_1(e_k) = \partial_0(e_\ell) = v$ for $k < \ell$. But then

$$e_1 e_2 e_3 \ldots e_k e_\ell \ldots e_n$$

is also a path from v_i to v_j. Either it is of length $< m$, or we can repeat the loop-removal process. Eventually, then, we reach a path of length $< m$ which runs from v_i to v_j. □

We can capture this in a formula as follows:

23 Corollary. *Let G be a directed graph with m vertices, and let C be its Boolean connection matrix. Form the sum of m Boolean matrices*

$$M = I + C + C^2 + \cdots + C^{m-1}.$$

Then there is a path from v_i to v_j in G if and only if $M_{ij} = 1$.

PROOF. By Theorem **21**, there is a path from v_i to v_j of length n just in case $C_{ij}^n = 1$. Thus

$$M_{ij} = I_{ij} \vee C_{ij} \vee C_{ij}^2 \vee \cdots C_{ij}^{m-1}$$

and so equals 1 if and only if there is a path from v_i to v_j of length n for some n with $0 \leq n \leq m - 1$. But this means, by Observation **22**, that $M_{ij} = 1$ if and only if there is a path from v_i to v_j of *any* length. □

24 Example. For the directed graph of Example **17**, $m = 3$ and so

$$M = I + C + C^2 = \begin{bmatrix} 1 & 0 & 0 \\ 0 & 1 & 0 \\ 0 & 0 & 1 \end{bmatrix} + \begin{bmatrix} 1 & 1 & 0 \\ 0 & 0 & 1 \\ 0 & 1 & 0 \end{bmatrix} + \begin{bmatrix} 1 & 1 & 1 \\ 0 & 1 & 0 \\ 0 & 0 & 1 \end{bmatrix} = \begin{bmatrix} 1 & 1 & 1 \\ 0 & 1 & 1 \\ 0 & 1 & 1 \end{bmatrix}$$

and M_{ij} does indeed equal 1 if and only if there is a path from vertex i to vertex j.

1. Let $(S, +, \cdot)$ be any semiring. Let $S^{n \times n}$ be the set of all $n \times n$ matrices with entries from S. Denote a typical element of $S^{n \times n}$ by (a_{ij}). Define addition on $S^{n \times n}$ by

$$(A + B)_{ij} = A_{ij} + B_{ij}$$

and define multiplication on $S^{n \times n}$ by

$$(A \cdot B)_{ij} = \sum_{k=1}^{n} A_{ik} \cdot B_{kj}$$

where the additions and multiplications on the right-hand side use the $+$ and $\cdot$, respectively, of $(S, +, \cdot)$. Show that:
   (i)   The $n \times n$ matrix 0 with $0_{ij} = 0$ for $1 \le i, j \le n$ is an identity for $+$ on $S^{n \times n}$.
   (ii) The $n \times n$ matrix I with $I_{ij} \equiv$ (**if** $i = j$ **then** 1 **else** 0) for $1 \le i, j \le n$ is an identity for $\cdot$ on $S^{n \times n}$.
   (iii) $(S^{n \times n}, +, 0)$ is a commutative semigroup.
   (iv) $(S^{n \times n}, \cdot, 1)$ is a semigroup.
   (v) $(S^{n \times n}, +, \cdot)$ is a semiring.

2. Given the following directed graph

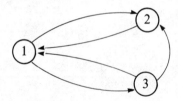

   (a) Compute the C, D, and K matrices for this graph.
   (b) Calculate the M matrix of Corollary 23 for this graph.

3. Recall the notion of a relation $R: A \to A$. We may represent it by a Boolean matrix M_R with rows and columns indexed by A, setting

$$(M_R)_{aa'} = \begin{cases} 1 & \text{if } aRa' \\ 0 & \text{if not.} \end{cases}$$

   Prove by induction that the nth power of M_R is the matrix of the relation R^n. Deduce that

$$M_{R^*} = \bigvee_{n \ge 0} M_R^n, \qquad M_{R^+} = \bigvee_{n > 0} M_R^n.$$

4. The matrix

$$C = \begin{pmatrix} 0 & 1 & 1 \\ 1 & 0 & 1 \\ 1 & 0 & 0 \end{pmatrix}$$

   is the Boolean connection matrix for what graph? Calculate C^2 and draw its corresponding graph.

6.3 Finite-State Acceptors and Their Graphs

In Section 2.3 we introduced the notion of a finite-state acceptor, a device for accepting sets of strings of symbols from a finite alphabet. Since FSAs have an underlying graph structure, our introduction to the formal machinery of graph theory earlier in this chapter will provide us with considerable insight into the behavior of these primitive machines. In this section, we first relate the reachability problem for such automata to the connectivity matrices of the last section. We then prove the equality between finite-state languages and the so-called regular languages, i.e., those which can be built up from finite subsets of X^* by a finite number of applications of the union, dot, and star operations of Section 2.2.

GRAPHS AND REACHABILITY

The *dynamics* (or *state-transition function*) of a finite-state acceptor is a map

$$\delta: Q \times X \to Q$$

where $Q = \{q_0, q_1, \ldots, q_n\}$ is the set of states and X is the set of inputs.

An FSA's state graph has Q for its set of nodes, and has one distinct edge from q_i to q_j for each x in X for which $\delta(q_i, x) = q_j$.

1 The *connection matrix* for this state-transition function is then the matrix A with

$$A_{ij} = \{x \in X \mid \delta(q_i, x) = q_j\}$$

Now recall from Section 2.3 the definition of the reachability function $\delta^*: Q \times X^* \to Q$ by

$$\delta^*(q, \Lambda) = q$$

$$\delta^*(q, wx) = \delta(\delta^*(q, w), x) \quad \text{for each } w \text{ in } X^* \text{ and } x \text{ in } X.$$

2 Definition. We say that q_j is *reachable* from q_i with respect to the dynamics δ just in case $q_j = \delta^*(q_i, w)$ for at least one string w in X^*.

3 Example. Consider the state graph

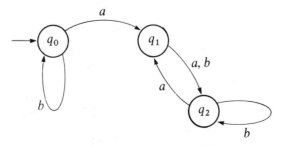

It has the connection matrix (with row and columns indexed 0, 1, 2)

$$A = \begin{bmatrix} \{b\} & \{a\} & \varnothing \\ \varnothing & \varnothing & \{a, b\} \\ \varnothing & \{a\} & \{b\} \end{bmatrix}.$$

We then compute

$$A^2 = \begin{bmatrix} \{bb\} & \{ba\} & \{aa, ab\} \\ \varnothing & \{aa, ba\} & \{ab, bb\} \\ \varnothing & \{ba\} & \{aa, ab, bb\} \end{bmatrix}.$$

We see that the ij entry of A^2 is just the set of paths w of length 2 for which $\delta^*(q_i, w) = q_j$.

We then have that

$$A^0 + A + A^2 = \begin{bmatrix} \{\Lambda, b, bb\} & \{a, ba\} & \{aa, ab\} \\ \varnothing & \{\Lambda, aa, ba\} & \{a, b, ab, bb\} \\ \varnothing & \{a, ba\} & \{\Lambda, b, aa, ab, bb\} \end{bmatrix}.$$

Note that q_j is reachable from q_i just in case the ij entry of $A^0 + A + A^2$ is nonempty.

In fact, we may apply Theorem **6.2.20** and Corollary **6.2.21** to deduce the following result:

4 Theorem. *Let* $\delta: Q \times X \to Q$ *have connection matrix* A. *Then the* (i, j) *entry of the* nth *power of* A, *for any* $n \geq 0$, *is the set of all length* n *paths from* q_i *to* q_j *in the state graph of* δ.

Moreover, if A *has* m *states, and we form the sum*

$$B = I + A + A^2 + \cdots + A^{m-1}$$

of X^*-matrices, *then* q_j *is reachable from* q_i *just in case* B_{ij} *is nonempty.*   □

REGULAR LANGUAGES

In Section 2.3, we introduced the notion of a *finite-state language* as being the set

$$T(M) = \{w \in X^* \mid \delta^*(q_0, w) \in F\}$$

of all strings accepted by some finite-state acceptor M. We suggested that such languages were closely related to those that could be built up using the union, dot, and star of Section 2.2.

5 Example. (In what follows, recall that all "missing transitions" lead to a single nonaccepting trap state.)

(i) Let M be given by

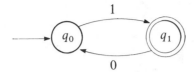

$T(M) = \{w \mid \delta^*(q_0, w) = q_1\}$, and we see that the only way to get from q_0 to q_1 is by applying a 1 to get to q_1 the first time, and then applying 01 any finite number $n \geq 0$ of times to traverse the loop to q_0 and back to q_1 again. Thus

$$T(M) = \{1\} \cdot \{01\}^*$$

which we simplify to $1(01)^*$.

(ii)

Here $T(M) = \{0, 1\}^* = (\{0\} \cup \{1\})^*$, which we may also denote by $(0 + 1)^*$.

(iii)

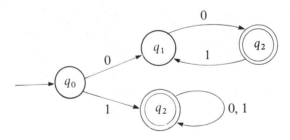

$$T(M) = 00(10)^* + 1(0 + 1)^*.$$

We devote the rest of this section to showing that any finite-state language can be represented in this fashion.

6 Definition. A subset of X^* is *regular* if it can be built up from finite subsets of X^* by a finite number of applications of $\cup$, $\cdot$, and $*$. We may also define the collection $\mathcal{R}(X)$ of regular sets on the alphabet X inductively as follows:

Basis Step: Each finite subset of X^* is in $\mathcal{R}(X)$.
Induction Step: If A and B are in $\mathcal{R}(X)$, then so too are $A \cup B$, $A \cdot B$, and A^*.

7 Theorem. *Every finite-state language is a regular set.*

PROOF. Given a finite-state acceptor $M = (Q, \delta, q_0, F)$ with input alphabet X, let

$$A_{qq'} = \{w \mid \delta^*(q, w) = q'\}$$

be the set of strings which send M from state q to state q'. Then

$$T(M) = \bigcup_{q \in F} A_{q_0 q}$$

so that we have proved our theorem if we can show that each $A_{qq'}$ is a regular set. To do this, we order the states Q of M as $\{q_0, q_1, \ldots, q_{n-1}\}$ and then set

$$A^s_{qq'} = \text{those strings which send } M \text{ from } q \text{ to } q' \text{ with only}$$
$$\{q_0, q_1, \ldots, q_{s-1}\} \text{ as intervening states.}$$

Thus the string $w = x_1 x_2 \ldots x_m$ belongs to $A^s_{qq'}$ just in case

$$\delta^*(q, x_1 x_2 \ldots x_j) \in \{q_0, q_1, \ldots, q_{s-1}\} \text{ for } 1 \leq j < m$$

while $\delta^*(q, x_1 x_2 \ldots x_m) = q'$.

We now prove by induction on s, $0 \leq s \leq n$, that each $A^s_{qq'}$ is a regular set — so that $A_{qq'} = A^n_{qq'}$ must also be regular for each pair of states q and q'.

Basis Step: For $s = 0$, the set of "intervening states" must be empty, and so

$$A^0_{qq'} = \{w \mid w \in X \cup \{\Lambda\} \text{ with } \delta^*(q, w) = q'\}$$

is a finite (possibly empty) set — and so is certainly regular.

Induction Step: Suppose that we already know that $A^s_{q_j q_k}$ is regular for some fixed value of s ($0 \leq s < n$) and for *any* pair of states q_j and q_k. We now prove that this implies that $A^{s+1}_{qq'}$ is also regular. To see this, note that a path from q to q' which only goes through intermediate states in $\{q_0, \ldots, q_s\}$ either goes only through $\{q_0, \ldots, q_{s-1}\}$

$$q \qquad\qquad q'$$

in which case the corresponding input sequence w is in $A^s_{qq'}$; or else the path goes through q_s at least once

$$q \qquad q_s \qquad q_s \qquad q_s \qquad \ldots \qquad q_s \qquad q'$$

so that the corresponding w belongs to $A^s_{qq_s} \cdot (A^s_{q_s q_s})^* \cdot A^s_{q_s q'}$.

Hence if each $A^s_{q_j q_k}$ is regular, then so too must be

$$A^{s+1}_{qq'} = A^s_{qq'} + A^s_{qq_s} \cdot (A^s_{q_s q_s})^* \cdot A^s_{q_s q'}.$$

We have thus proved by induction that each $A_{qq'} = A_{qq'}^n$ is regular, and hence $T(M) = \bigcup_{q \in F} A_{q_0 q}$ is also regular.   □

We have just proved that every finite-state language is a regular set. Our goal in the rest of the section is to prove the converse — that every regular set is a finite-state language. Now we already know (**2.3.6**) that every finite set is a finite-state language. So we only have to prove that if A and B are both finite-state languages, then so too are $A \cup B$, $A \cdot B$, and A^*.

8 Example. The language $(\{a^4\} \cup \{a^6\})^*$ can be recognized by the FSA

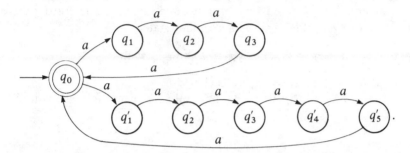

But there is a problem! There are two different arcs labeled "a" emanating from q_0, one leading to the a^4 loop, the other leading to the a^6 loop. According to the definition of $\delta: Q \times X \rightarrow X$, such a device is illegal. To permit transitions of the type indicated, δ would have to be redefined as a map

$$\delta: Q \times X \rightarrow 2^Q$$

sending a state/symbol pair to a set of states.

9 Definition. A *nondeterministic finite-state acceptor* (*NDA*) M with input alphabet X is specified by a quadruple (Q, δ, Q_0, F) where

Q is a finite set of states
$\delta: Q \times X \rightarrow 2^Q$ assigns to each state/input pair (q, x) a set $\delta(q, x) \subset Q$ of *possible* next states.
$Q_0 \subset Q$ is the set of initial states
$F \subset Q$ is the set of accepting states.

We shall see that an NDA is just like an FSA except that there is a *set* of initial states and a *set* of possible next states. We can motivate this *nondeterminism* — the fact that at any stage of processsing there is a *set* of possible states — by briefly considering the problem of *parsing* a sentence of English, that is, assigning to the sentence its grammatical structure. Consider a machine which reads in one word at a time of a sentence, and considers all possible analyses consistent with what it has seen before. After reading in

"The red wood"

it could be in either of two states

<blockquote>
"determiner adjective adjective"; or
</blockquote>

<blockquote>
"determiner adjective noun."
</blockquote>

The first would work if the completion of the sentence were "The red wood box is heavy," while the second would work with the sentence "The red wood is beautiful" — but there is no way that either can be completed in a way that will let the parsing machine construe "The red wood gracefully" as a grammatical English sentence.

This motivates considering all possible states to which an input string w can lead an NDA M from its initial states, and saying w is accepted if there is *at least* one way of getting to an accepting state.

10 We extend $\delta\colon Q \times X \to 2^Q$ to $\delta^*\colon 2^Q \times X^* \to 2^Q$ by:

Basic Step: $\delta^*(p, \Lambda) = \{p\}$ for each $p \subset Q$.
Induction Step: $\delta^*(p, wx) = \bigcup \{\delta(q, x) \mid q \in \delta^*(p, w)\}$ for each w in X^*, x in X, $p \subset Q$.

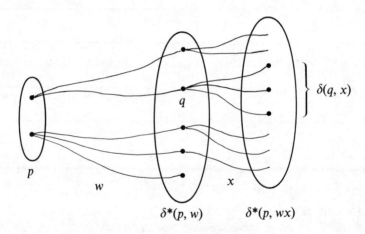

We then set

$$T(M) = \{w \mid w \text{ in } X^* \text{ and } \delta^*(Q_0, w) \cap F \neq \varnothing\}.$$

The reader should have little trouble completing the proof of the following:

11 Proposition. *A subset of X^* is a finite-state language if and only if it is a $T(M)$ for some NDA M.*

PROOF OUTLINE. (*i*) If $L \subset X^*$ is a finite-state language, it is $T(M_1)$ for some FSA $M_1 = (Q_1, \delta_1, q_1, F_1)$. We define an NDA $M_2 = (Q_1, \delta_2, \{q_1\}, F_1)$ in terms of M_1 by setting $\delta_2\colon Q_1 \times X \to 2^{Q_1}\colon (q, x) \mapsto \{\delta_1(q, x)\}$. Then (check the details) $T(M_1) = T(M_2)$.

(ii)   If $L \subset X^*$ is $T(M)$ for the NDA $M = (Q, \delta, Q_0, F)$ we may define an FSA $M' = (Q', \delta', q', F')$ in terms of M by

$$Q' = 2^Q$$

$$\delta'(p, x) = \bigcup\{\delta(q, x)|q \in p\} \quad \text{for each } p \text{ in } Q' \text{ and } x \text{ in } X$$

$$q' = Q_0 \in Q'$$

$$F' = \{p|p \cap F \neq \varnothing\} \subset Q'.$$

Then (check the details) $T(M) = T(M')$.　　　　　　　□

12 Example. Consider the NDA

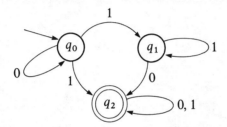

which accepts the language $0^* \cdot 1 \cdot (0 + 1)^* + 0^* \cdot 1 \cdot 1^*0 \cdot (0 + 1)^*$. The two factors in this expression correspond to the two paths from q_0 to q_2.

The FSA corresponding to this device has 8 states:

$$\varnothing, \{q_0\}, \{q_1\}, \{q_2\}, \{q_0, q_1\}, \{q_0, q_2\}, \{q_1, q_2\}, \{q_0, q_1, q_2\}.$$

Its start state is $\{q_0\}$, and its final states are $F = \{\{q_2\}, \{q_1, q_2\}, \{q_0, q_2\}, \{q_0, q_1, q_2\}\}$. Finally, its transition function δ is given by the following chart.

STATE	INPUT	
	0	1
$\{q_0\}$	$\{q_0\}$	$\{q_1, q_2\}$
$\{q_1\}$	$\{q_2\}$	$\{q_1\}$
$\{q_2\}$	$\{q_2\}$	$\{q_2\}$
$\{q_0, q_1\}$	$\{q_0, q_2\}$	$\{q_1, q_2\}$
$\{q_0, q_2\}$	$\{q_0, q_2\}$	$\{q_1, q_2\}$
$\{q_1, q_2\}$	$\{q_2\}$	$\{q_1, q_2\}$
$\{q_0, q_1, q_2\}$	$\{q_0, q_2\}$	$\{q_1, q_2\}$

To complete our proof that every regular set is a finite-state language, we need only prove the following:

13 Theorem. *Let M_A and M_B be NDAs which accept the sets $A = T(M_A)$ and $B = T(M_B)$ respectively. Then we can define NDAs $M_{A \cup B}$, $M_{A \cdot B}$, and M_{A^*} which accept $A \cup B$, $A \cdot B$ and A^*, respectively.*

PROOF. Let

$$M_A = (Q_A, \delta_A, Q_{0A}, F_A)$$

$$M_B = (Q_B, \delta_B, Q_{0B}, F_B).$$

(i)   We set (using $+$ for disjoint union)

$$M_{A \cup B} = (Q_A + Q_B, \delta_{A \cup B}, Q_{0A} + Q_{0B}, F_A + F_B),$$

where

$$\delta_{A \cup B}(q, x) = \begin{cases} \delta_A(q, x) & \text{if } q \in Q_A, \\ \delta_B(q, x) & \text{if } q \in Q_B. \end{cases}$$

Then

$$T(M_{A \cup B}) = \{w \mid \delta^*_{A \cup B}(q, w) \cap (F_A + F_B) \neq \varnothing \text{ for } q \in Q_{0A} + Q_{0B}\}$$

$$= \{w \mid \delta^*_A(q, w) \cap F_A \neq \varnothing \text{ for } q \in Q_{0A}\}$$

$$\cup \{w \mid \delta^*_B(q, w) \cap F_B \neq \varnothing \text{ for } q \in Q_{0B}\}$$

$$= A \cup B.$$

(ii) $A \cdot B$:

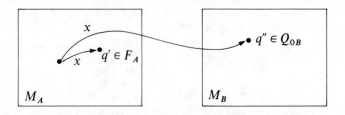

The idea of the construction is that M_A should continually monitor a string w — but now instead of simply entering an accepting state in F_A when an initial segment belongs to A, it must also have M_B enter its initial states Q_{0B} (in addition to whatever states it may already be in — remember that M_B is nondeterministic), so that it may also check if the remainder of the string belongs to B:

$$M_{A \cdot B} = (Q_A + Q_B, \delta_{A \cdot B}, Q_{0A}, F_B)$$

where

$$\delta_{A \cdot B}(q, x) = \begin{cases} \delta_A(q, x) & \text{if } q \in A \text{ and } \delta_A(q, x) \cap F_A = \emptyset, \\ \delta_A(q, x) \cup Q_{0B} & \text{if } q \in A \text{ and } \delta_A(q, x) \cap F_A \neq \emptyset, \\ \delta_B(q, x) & \text{if } q \in B. \end{cases}$$

(iii) A^*

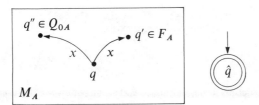

The same idea applies for A^*: M_A must be modified so that whenever it has checked that the string so far belongs to A^*, it must not only continue checking but must also reactivate its initial states.

$$M_{A^*} = (Q', \delta_{A^*}, Q'_{0A}, F'_A)$$

where $Q' = Q_A, Q'_{0A} = Q_{0A}$, and $F'_A = F_A$ if $F_A \cap Q_{0A} \neq \emptyset$ but $Q' = Q_A \cup \{\hat{q}\}$ $Q'_{0A} = Q_{0A} \cup \{q'\}$ and $F'_A = F_A \cup \{\hat{q}\}$ for a new state $\hat{q}$, otherwise. This makes sure that the empty string Λ will be accepted in A^*.

$\delta_{A^*}(q, x)$ is only defined for $q \in Q_A$ and satisfies

$$\delta_{A^*}(q, x) = \begin{cases} \delta_A(q, x) & \text{if } \delta_A(q, x) \cap F_A = \emptyset, \\ \delta_A(q, x) \cup Q_{A0} & \text{if } \delta_A(q, x) \cap F_A \neq \emptyset. \end{cases} \qquad \square$$

14 Example. $A = 01^*$ is accepted by the NDA

Now $A^* = (01^*)^*$, which equals $\{\Lambda\} + 0 \cdot (0 + 1)^*$. Let us check that this is indeed accepted by M_{A^*} which, according to the above construction, equals

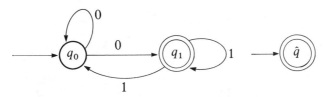

i.e., we make $\hat{q}$ an initial as well as accepting state, and we add a transition to q_0 in M_{A^*} for each transition to an accepting state (i.e., q_1) in M_A.

Since $\hat{q}$ is both a start and final state of M_{A^*}, we see that $\Lambda \in T(M_{A^*})$. Since $\delta_{A^*}(q_0, 1) = \varnothing$, we see that no string starting with 1 can belong to $T(M_{A^*})$. Now

$$\delta_{A^*}(q_0, 0) = \{q_0, q_1\}.$$

$$\delta_{A^*}(\{q_0, q_1\}, 0) = \delta_{A^*}(q_0, 0) \cup \delta_{A^*}(q_1, 0) = \{q_0, q_1\} \cup \varnothing = \{q_0, q_1\}.$$

$$\delta_{A^*}(\{q_0, q_1\}, 1) = \delta_{A^*}(q_0, 1) \cup \delta_{A^*}(q_1, 1) = \varnothing \cup \{q_0, q_1\} = \{q_0, q_1\}.$$

Thus $\delta_{A^*}(q_0, w) = \{q_0, q_1\}$ for every w in $0 \cdot (0 + 1)^*$. Thus $T(M_{A^*}) = \{\Lambda\} \cup 0 \cdot (0 + 1)^* = A^*$, as was to be shown.

EXERCISES FOR SECTION 6.3

1. Use the construction described in the text to provide the state diagram of the FSA M_L which accepts the sets:
   (a) $L = \{0, 1, 001, 1110, 1111\}$
   (b) $L = \{\Lambda, 0, 1, 00, 11111\}$.

2. Show that $(0 + 1)^* \neq 0^* 1^*$.

3. Construct NDAs which accept the languages $A = (0 + 1)^* \cdot 01$ and $B = 01^* + 1(1 + 00)^*$. Then use the constructions given in the text to provide NDAs which accept $A \cup B$, $A \cdot B$ and A^*, respectively.

4. Supply the formal details of the proof that a subset of X^* is a finite-state language if and only if it is accepted by some NDA.

5. Let X be a finite alphabet with $L \subset X^*$. Define

$$\text{Prefix}(L) = \{w \in \Sigma^* \,|\, \text{for some } z \in \Sigma^*, wz \in L\}.$$

Prove that if L is a finite-state language, then so too is Prefix (L). (Hint: Relate the prefix relationship on strings to the reachability condition on the state graph of an FSA.)

6. Describe an NDA that accepts the language over the alphabet $\Sigma = \{0, 1\}$ generated by the following productions: $S \to 0A$, $S \to 1$, $A \to 0A$, $A \to 1$, $A \to 1S$. (Hint: Use the nonterminal letters S and A as names for states.)

Author Index

Notation Index

Subject Index

Texts and Monographs in Computer Science